Dinosaurs and Dioramas

Dinosaurs and Dioramas

CREATING NATURAL HISTORY EXHIBITIONS

Sarah J. Chicone and Richard A. Kissel

Left Coast Press Inc.

Walnut Creek, CA

Left Coast Press, Inc.
1630 North Main Street, #400
Walnut Creek, CA 94596
www.LCoastPress.com

ISBN 978-1-61132-274-3 hardback
ISBN 978-1-61132-275-0 paperback
ISBN 978-1-61132-276-7 institutional eBook
ISBN 978-1-61132-277-4 consumer eBook

Library of Congress Cataloging-in-Publication Data
Chicone, Sarah J.
Dinosaurs and dioramas : creating natural history exhibitions /
Sarah J. Chicone and Richard A. Kissel.
pages cm
Summary: "Two experienced exhibit designers lead you through the complex process of design and installation of natural history exhibitions. The authors introduce the history and function of natural history museums and their importance in teaching visitors the basic principles of science. The book then offers you practical tricks and tips of the trade, to allow museums, aquaria, and zoos--large or small--to tell the story of nature and science. From overall concept to design, construction, and evaluation, the book carries you through the process step-by-step, with emphasis on the importance of collaboration and teamwork for a successful installation. A crucial addition to the bookshelf of anyone involved in exhibit design or natural history museums"-- Provided by publisher.
Includes bibliographical references and index.
ISBN 978-1-61132-274-3 (hardback) -- ISBN 978-1-61132-275-0
(paperback) -- ISBN 978-1-61132-276-7 (institutional ebook)
--ISBN 978-1-61132-2774 (consumer eBook)
1. Natural history museums--Educational aspects. 2. Museum
exhibits. I. Kissel, Richard A. II. Title.
QH70.A1C49 2013
508.075--dc23
2013025286

Printed in the United States of America
∞ ™ The paper used in this publication meets the minimum requirements of American National Standard for Information Sciences—Permanence of Paper for Printed Library Materials, ANSI/NISO Z39.48–1992.

On the cover: Skeleton of the dinosaur *Allosaurus fragilis* (Copyright Joshua Franzos for Carnegie Museum of Natural History), top, and diorama of Bighorn Sheep, *Ovis canadensis* (Photo by Kelly Kennard, Courtesy Yale Peabody Museum of Natural History).

Cover and text design by Lisa Devenish, Devenish Design

CONTENTS

Preface

The stories of natural history are worth celebrating. They are the stories of ancient beasts that thundered across primitive landscapes, they are the stories of exotic flora from far-away lands, and they are the stories of human culture, both foreign and familiar. All of nature is related and intertwined, and for this reason, we are but a single part of its endless complexity. Across millions of years and untold generations, our lineage has evolved to value and appreciate nature. We internalize and release individualized interpretations through art, and our boundless curiosity leads to questions and the desire to explain what we see—science. Those museums that present the stories of nature—from zoos and aquaria to natural history museums and nature centers—find themselves at the exact intersection of art and science.

The chapters contained within the following pages focus on the creation of science-based exhibitions[1] within natural history settings. Organizationally, *Dinosaurs & Dioramas* is divided into two distinct sections. The first introduces the history of natural history display, and it also includes an in-depth discussion on the nature and presentation of science. The second and larger section takes readers on a journey through the process of exhibition development, from topics of visitor engagement to strategies of writing and design.

As highlighted within the text, today's museums find themselves in an interesting position. With concerns about the planet's health—as well as that of the millions of species that call it home—dominating conversation, scientific literacy within society is increasingly important. And with the average American spending less than five percent of his or her life in the classroom (Falk and Dierking 2010), the role of informal education within our society has never been as important as it is this moment. It is the local institution of nature and science that can help a growing population not only understand the Earth but also help preserve it and make educated decisions regarding its future and the sustainability of its resources. Scientists and artists—with the exhibitions that they create *together*—have the power to inspire. We are all teachers. The chapters herein outline the process of exhibition development

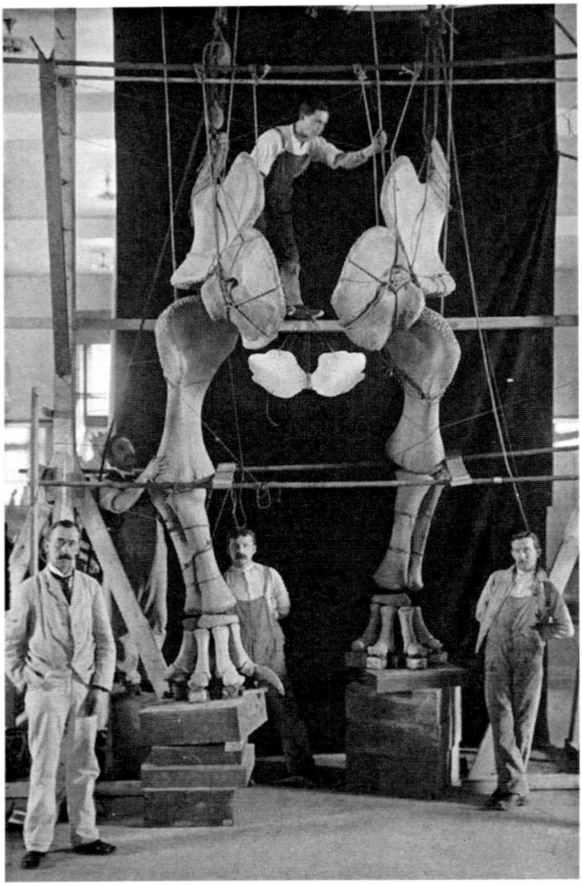

The mounting of a "*Brontosaurus*" skeleton in 1904 at the American Museum of Natural History. The skeletons of these giant creatures have been inspiring museum visitors for well over a hundred years. Image #17506 American Museum of Natural History Library.

and design. Coupled with your passion, commitment to education, and respect for the museum visitor, they can lead to more than just an opening day; they can lead to exhibitions that successfully engage visitors and change lives.

In writing this text, it is our sincere hope that its words will provide both museum professionals and students with an a-to-z guide on the creation of an exhibition, as well as discussions regarding the importance of natural history museums within the 21st century. More importantly, we hope that the book sparks conversation among the many talents that populate the museum field. It is written for the graphic designer and the developer; it is written for the model-maker and the curator.

Developing an exhibition, from its initial inception to its public opening, is an amazing process of creativity, artistry, and learning. With *Dinosaurs and Dioramas*, let's together explore, embrace, and celebrate the many aspects of natural history exhibitions and the inherent responsibility associated with their creation.

Acknowledgments

For those dedicated to inspiring the scientist in us all...

While the words within these pages are ours, they were written with the support of many. We thank our editors Mitch Allen and Stefania Van Dyke for their wonderful guidance and support, as well as their unwavering trust. For reviewing an early draft of the text, we are indebted to Erica Kelly and David Whitemyer; their comments greatly improved the manuscript, and their encouraging words helped more than they know. A special thank you to our colleagues Paul Pearson and Liza Rawson, talented exhibition developers and educators who contributed both their knowledge and expertise to *Exhibition Strategies*, a graduate course in the Johns Hopkins University's Museum Studies graduate program; this course and our many wonderful students are the inspiration behind this volume. We would also like to extend thanks to our friend and colleague Phyllis Hecht for her always-endless support. Finally, in addition to those researchers and authors cited throughout the text, we thank all of those within the museum field. Your dedication to educating and inspiring visitors around the world is nothing short of remarkable.

From Sarah...

My work would not have been possible without the loving support of my entire family—Jim and Karen Morby, David and Jen Chicone, and Doug and Deborah DePuy. It does indeed take a village. A special thank you to my husband Vincent for continually encouraging my passions and keeping me grounded. To my darling daughters Ella and Loretta, nothing compares to the awe, excitement, and curiosity about the world I see reflected in your eyes. I will always be inspired by your thirst for knowledge and the simple joys you find in our natural world. Thank you both for reminding me to slow down and look... To all those adults who encouraged my seven-year-old self and opened my eyes to the glories of the natural history museum, I am forever grateful.

I have been lucky enough to cross paths with many fabulous exhibitions and museum folks over the years, and I have learned much from all of you who "do" (a special thank you to Phyllis, Karen, Judy, Deborah, Liza, and Paul). My dear friend Dr. Kissel, thank you for joining me on this journey; it's been quite a ride!

From Richard...

For her love and support, I would like to thank my amazing wife Kelly. I am also forever grateful for my family's encouragement and support throughout the years. Those many trips to the *Dinosaur Hall* at Pittsburgh's Carnegie Museum led to the life that I lead today. And while that old hall is now gone—a casualty of changing science and resulting renovation—it will forever remain my favorite parade of skeletons. Throughout the years since, I have met and learned from too many to list, but am especially grateful to my scientific mentors Mary Dawson, Don Steinker, Thomas Lehman, Hans Sues, and Robert Reisz, and to those colleagues with whom I've shared so many great experiences developing exhibitions, from Beth and Leon to Todd, Gretchen, A-Pat-o, and Steve. And especially Erica; I will always be your student. Last but certainly not least, I'd like to thank my kitten Pria, who spent her evenings by my side, watching me write while basking under the warm glow of my desk lamp. And thank you, Dr. Chicone; it's been a pleasure!

Part 1

Science Sets the Stage

A Brief History of Natural History Museums

"Nature is so powerful, so strong. Capturing its essence is not easy..."

—Annie Leibovitz

Natural history collections are the foundation for our understanding of the natural world. Whether it's a massive skeleton of an *Apatosaurus* or the delicate carvings of the Rosetta Stone, the objects of natural history inspire wonder and awe, and they provide humanity with a sense of place within our world. The human species is embedded in nature; we share in the "history of all natural living beings" (Cuvier 1797 in Taquet 2007, 4), and the role of natural history museums is to present that story to all. As gateways to inspiration and knowledge, museum exhibitions are critical to society's advancement and perception of itself. For this reason, exhibition developers and designers are both communicators and teachers; their influence is far-reaching and they shoulder tremendous responsibility as a result.

This text is intended to help you navigate the design and development of natural history exhibitions; it offers a consideration of scientific process, visitor experience, and practical tricks and tips of the trade. This is a pragmatic guide tied specifically to those institutions dedicated to telling the story of nature and the science by which we know it—from aquaria and zoos to nature centers and natural history museums. We begin our journey by placing these venerable institutions in historical context.

From Private Collections to Public Institutions: Asserting Control over the Natural World

During the Middle Ages, there was little interest in natural history; collections did exist but were confined to wealthy abbeys and cloisters (Hagen 2008). The European Renaissance, or "re-birth," followed, set in motion by scholars and artists who traced their roots to the classical world. This movement represented an intellectual attempt to align with ancient wisdom and tradition, essentially relocating learning to the grove of the muses—the nine goddesses of poetry, music, and the liberal arts (Findlen 2003). With an influx of new specimens and artifacts,[1] preservation techniques, exploration, and religious reformation, there was a need to explain and order an increasingly illogical and pluralistic world, and a need to collect and arrange knowledge in an attempt to control it. The museum, rooted in scholarship, was—as some suggest—the "most complete response to the crisis of knowledge provoked by the expansion of the natural world through the voyage of discovery and exploration..." (ibid., 36). As collections continued to grow, the sheer number of *things* increased the need and desire to find order among them.

Our story begins, as many anthologies of natural history do, with the ubiquitous cabinets of curiosity known as Kunst- und Wunderkammers, studiolos, gallerias, or museos (Koeppe 2002). These were rooms (kammer) of art (kunst) and marvels (wunder), literally cabinets of "remarkable" things that, when taken together, reflected the authors' view of the world. They were "encyclopedic collections of all kinds of objects of dissimilar origin and diverse materials on a universal scale" (ibid., para 1)—collections of natural and artificial objects intended to represent the world in miniature (Fig. 1.1). Especially popular among the wealthy merchants and nobility of Europe, these early analogs often straddled the public and private divide (Findlen 2003). Some were kept under lock and key, closed sanctuaries for their owners—like Francesco I de'Medici's sixteenth-century studiolo in Florence's Palazzo Vecchio, a private collection and microcosm of his understanding and command of the world (Kinch 2011). Others were displays of wealth and self-aggrandizement, showcases for other prosperous and landed nobility—like Holy Roman Emperor Rudolf II's Kunstkammer at the Hradschin Palace in Prague (Evans 1973). These collections of art, nature, and human engineering—large and small, shared or private—represent the precursor of the modern natural history museum.

An Affair to Remember: Science and Natural History

But how did these first cabinets of curiosity, with everything from "unicorn" horns to parades of shells, evolve into to the natural history museums of today—public institutions that promote scientific understanding and knowledge? The eighteenth century witnessed a number of significant milestones that impacted the development and institutionalization of natural history museums. As noted by Hoquet (2010, 31), "'natural history' was understood as the science of '*naturalia*'—that is, the study of the three kingdoms of nature (mineral, vegetable, animal)." This broad and encompassing idea of natural history was based on Pliny the Elder's *Historia naturalis* (circa AD 77), which included 37 books that examined everything from beehives to the universe (Hoquet 2010). The 'universal history' of the Plinian model was reflected in leading eighteenth-century naturalist Georges-Louis Leclerc, Comte

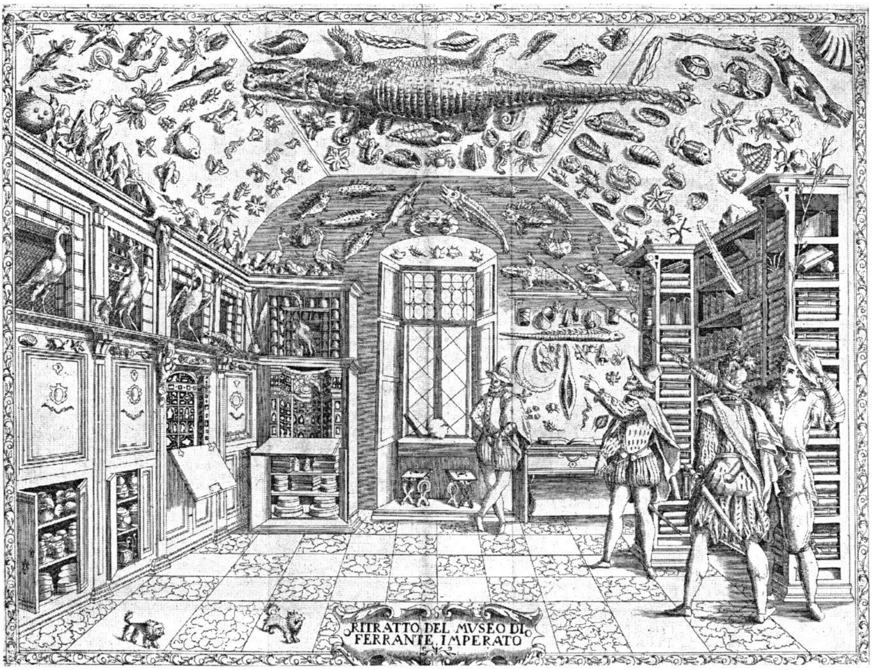

Figure 1.1: An engraving of the collection of Ferrante Imperato—a Neapolitan apothecary—at the Palazzo Orsini di Gravina represents the earliest known pictorial representation of a natural history cabinet. Public domain, Wikimedia Commons.

de Buffon's popular thirty six-volume *Histoire naturelle*, published between 1749 and 1789. Buffon's published work, along with his involvement in the transformation and expansion of the Jardin du Roi public garden in Paris (now the Jardin des Plantes), worked to define "natural history as a scientific enterprise" (Spary 2000, 16).

In addition to Buffon's contributions, Swedish botanist, zoologist, and physician Carl Linnaeus' *Systema Naturae* (1735) and *Species Plantarum* (1753) had a direct impact on the ordering of knowledge. Prior to Linnaeus, there existed no consistent practice for scientifically naming species. Students of nature often used long, unwieldy Latin names, and many species received different names by different scientists, making both communication and any systematic organization of nature difficult. The wild briar rose, for example, had been named both *Rosa sylvestris inodora seu canina* ("odorless woodland dog rose") and *Rosa sylvestris alba cum rubore, folio glabro* ("pinkish white woodland rose with hairless leaves"). But throughout his works, Linnaeus developed the naming system still used today: binomial nomenclature. In this system, the names of species consist of two names, such as *Tyrannosaurus rex* or *Homo sapiens*, and species are classified within a hierarchical structure (Fig. 1.2).

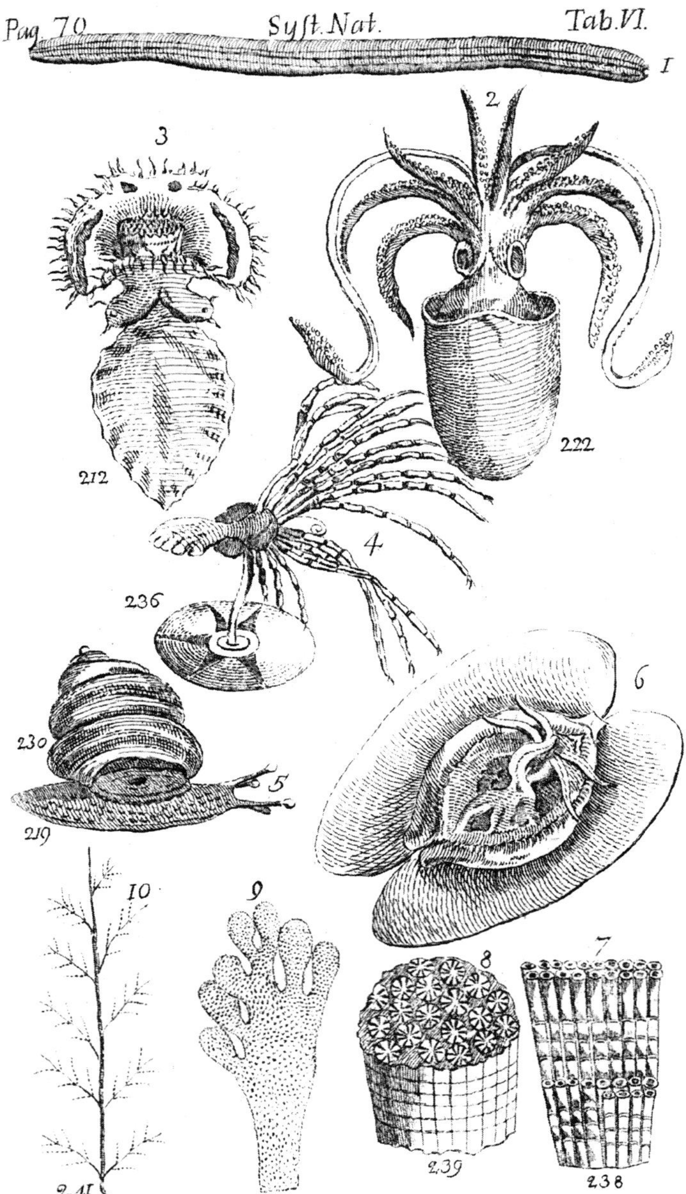

FIGURE I.2. Image from *Systema Naturae*, highlighting invertebrate animals within Linnaeus' Classis Vermes. The tenth edition of his *Systema Naturae*, published in two volumes in1758 and 1759 is considered the beginning of zoological nomenclature. Linnaeus' hierarchy classified living things within a simple, linear organization of most general (Imperium) to most similar (Varietas). This system evolved into that still used by many biologists today—a nested series of ranks: (Kingdom (Phylum (Class (Order (Family (Genus (Species))))))), though some recent authors have proposed yet another evolution of the system (e.g., de Queiroz and Gauthier 1992). Public domain, Wikimedia Commons.

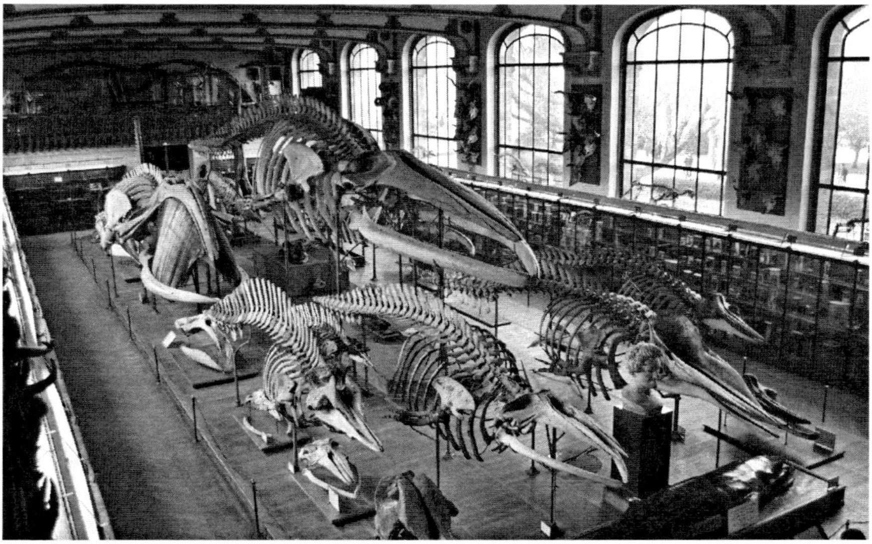

FIGURE 1.3. Today, Cuvier's collections can be seen in the Galeries de Paléontologie et d'Anatomie comparée (Galleries of Palaeontology and Comparative Anatomy), part of the French Muséum national d'Histoire naturelle (National Museum of Natural History). Steren Giannini, Pannini, GFDL, Wikimedia Commons.

Linnaeus' work imposed order upon nature, and it was later described by Hagan (2008, 45) as the "'printed instructions' for the arrangement of a museum." As Hagan claimed, "The clear and logical mind of Linnaeus not only purified the system, but also enabled him to purge the collections of a considerable number of fabulous and fictitious objects..." (ibid., 45); indeed, much to the chagrin of the seven-year-old in all of us, this moment is when we lost the unicorn horns and the dragon's blood.

Despite their tenuous personal relationship and their differences in understandings of species and their classification[2] (Sloan 1976; Conniff 2006, 2008), Buffon and Linnaeus, as well as other early advocates of natural history, had a significant impact on the way natural objects were conceived of and arranged—in relation to man (as in Buffon) or to each other (as in Linnaeus)—as well as the way relationships within the natural world were understood.

Early methods of classification were later expanded through the comparative anatomy studies of French anatomist Georges Cuvier's Cabinet d'anatomie comparée in Paris (Fig. 1.3). In 1802, when Cuvier became the Chair of Animal Anatomy at the Muséum national d'Histoire naturelle—formerly Buffon's Cabinet d'Histoire naturelle in the Jardin du Roi—he began an aggressive expansion of the collections that lasted until his death in 1832. During his tenure the collection grew nearly 600 percent, from 2,898 preparations in 1802 to some 16,665 preparations in 1832 (Taquet 2007). This was a site of knowledge development and, ultimately, his museum "became the paradigm for museums around the world. They too wished to adopt a view of the museum as active experiment rather than passive representation" (ibid., 13).

A Triad of Purposes: Preservation, Research, and Education

It was this idea—that active scientific research should be located within the institution that also promoted the curation and display of collections—that saw museums and the object-based epistemology championed during the nineteenth century as central to the creation of disciplinary boundaries within the natural and social sciences, including archaeology, geology, paleontology, and biology (Conn 1998; Knell 2007). Conn (1998, 15) suggests that objects during the late Victorian age were seen as "the sites of meaning and knowledge," and, given that museums were the repositories of such objects, "many intellectuals regarded museums as a primary place where new knowledge about the world could be created and given order." Because scientific collections were expensive to collect and curate, beyond the reach of most liberal arts colleges and aspiring universities, "Natural history museums were the principal location for dialogues and the exchange of specimens among those debating the identification and connection among natural objects (and, later, human artifacts)" (Kohlstedt 1995, 151). Historians like Mary Winsor suggest that natural history museums, such as Louis Agassiz's Museum of Comparative Zoology at Harvard University, were "scientific instruments, tools for investigation in the quest to understand living things" (ibid., 159).

An important aspect of museums as sites of knowledge creation was the subsequent availability of this knowledge to the "masses" in the form of exhibitions and programming; meaning was ultimately created and given form through the way objects were presented in displays (Conn 1998). Ichthyologist G. Brown Goode, Assistant Secretary of the Smithsonian Institution from 1887 to 1896, "publicized the central triad of purposes that marked urban museums at the end of the [19th] century—preservation, research, and education—and he developed a rationale for display that emphasized the importance of linking research ideas with the organized exhibitions of selected materials, including those of science and technology" (Kohlstedt 1995, 155).

The "Golden Age" of the American natural history museum, which reached its peak in the 1920s and 30s (Rader and Cain 2008), saw the popularization and institutionalization of classic natural history exhibitions. In 1927, Gladwyn Kingsley Nobel—head of the American Museum of Natural History's Department of Animal Behavior—insisted that "curators must...design exhibits that 'dissect and analyze Nature in such a way that the public will understand the principles controlling the life of the creatures portrayed'" (Nobel in Rader and Cain 2008, 157). The artistic efforts of Carl Akeley, noted for his realistic and taxidermal displays at the American Museum of Natural History, and artist Charles R. Knight (Fig 1.4) brought nature to life and popularized understandings of biological and anthropological development over time (Kohlstedt 1995, 160).

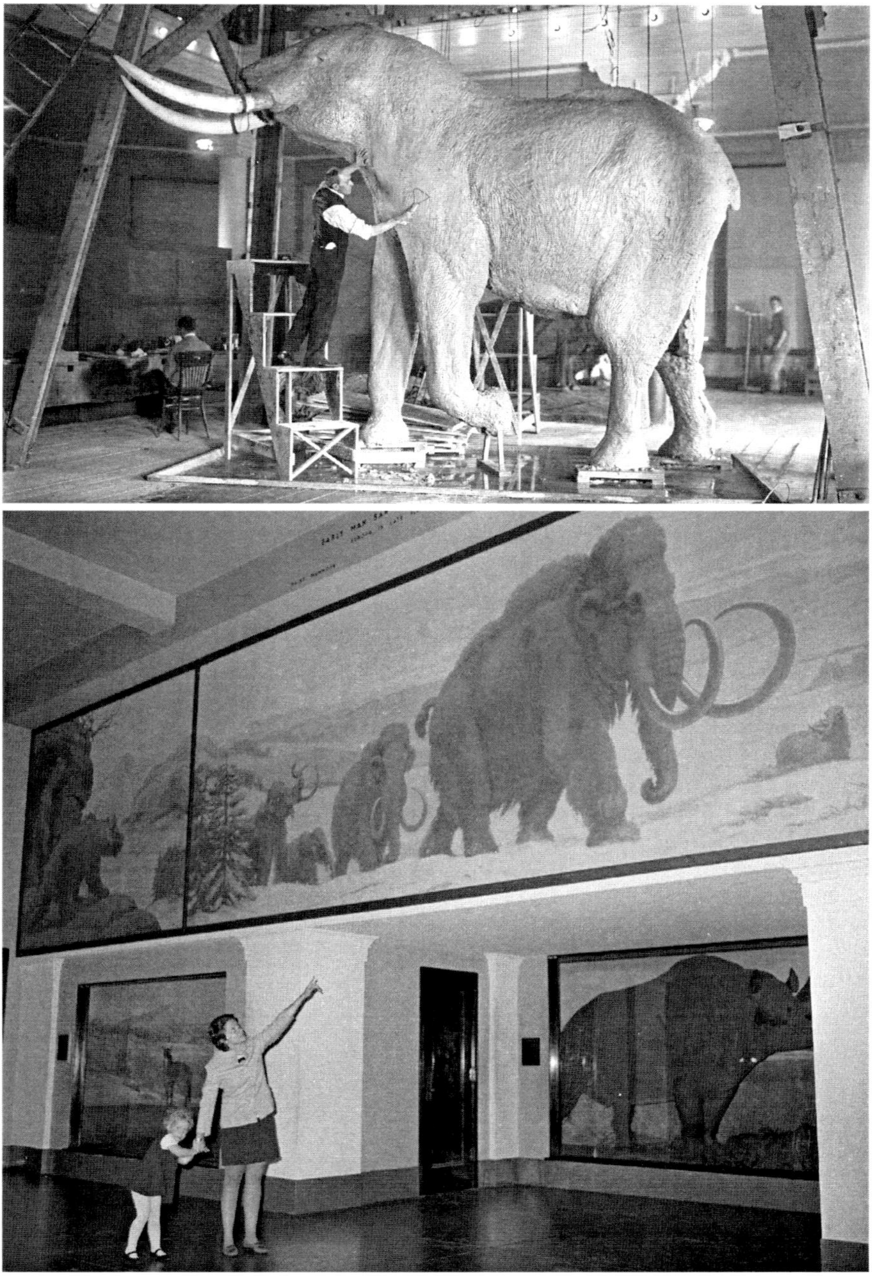

FIGURE 1.4. Carl Akeley modeling an elephant in 1914 for the American Museum of Natural History's African Hall (top). In the 1920s, Charles R. Knight painted a series of 28 murals for the fossil halls of The Field Museum, bringing long-extinct beasts and more than two billion years of Earth's history to life; shown here are Knight's granddaughter Rhoda and her daughter Melissa admiring his mammoths and cave bear in 1969 (bottom). (top) Image # 34314 American Museum of Natural History Library; (bottom) © The Field Museum, #GN81615_2.

As attention continued to shift toward education during the 20th century, Rader and Cain (2008, 155) suggest that the "three part mission of research, education, and exhibition tilted rapidly toward exhibition," and it became a question of how to balance "educational goals with scientific ones." There existed a general feeling that commitment to exhibition—both in the form of staff time and resources—had compromised commitment to scientific research. And by the mid-20th century, exhibition departments were gaining ground within natural history museums, reflecting a "broader professionalization of museum exhibition... taking place nationwide" (159). Despite the strides in the professionalization of exhibition and visitor engagement, research staff continued to malign what they saw as a direct challenge to scientific scholarship, what some vocal critics called the "museums as Disneyland" approach (Terrell 1991, 152).

Natural History: The Way of the Dodo?

By the 1960s, fresh off the heels of Sputnik during the height of the Cold War, United States policy makers were an increasingly important constituency for many natural history museums, which found themselves in a new crisis: How best to create exhibitions that presented not only the topics of science but also best embodied new educational ideals? As Rader and Cain (2008, 163) suggest, "Exhibits had become a new battleground for both the definition of science itself, as well as how museums (as institutions of science) might best promote public engagement with it." There was a shift from a focus on specimens and their broader relationship to the natural world—the backbone of natural history—to "abstract life science principles" (ibid., 163-165), a "gradual shift from specimen-display to non-collection exhibits—from natural history to science." During the 1970s, natural history museums were confronted with the role their collections and specimens played in defining their institutions and its future (Rader and Cain 2008).

Today, many institutions have returned to object-based exhibitions. In 2006, for example, Chicago's Field Museum unveiled *Evolving Planet*, a 27,000-square-foot, permanent exhibition on the history of life on Earth. Though the footprint was similar to that of its predecessor—*Life Over Time*—the renovated galleries contained nearly twice as many objects. And the Muséum national d'Histoire naturelle's *Grande Galerie de l'Évolution* harkens back to Cuvier's specimen-rich hall. The objects, after all, are and will forever remain the basis of humanity's understanding of nature; their position at the forefront of any natural history display reflects and reinforces that importance.

In February 2012, the Smithsonian National Museum of Natural History hosted a workshop entitled *21st Century Learning in Natural History Settings*. There, staff from 100 institutions—from museums and zoos to aquaria, science centers, and nature centers—discussed the future role of

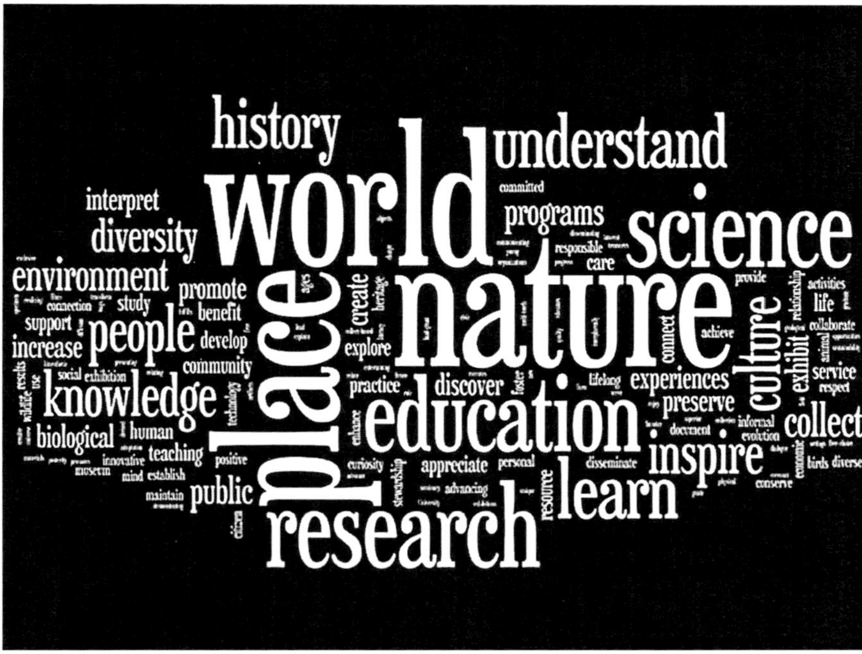

FIGURE 1.5. Word cloud generated from the mission statements of the 100 institutions participating in the 2012 *21st Century Learning in Natural History Settings* conference. The terms "Earth" and "world" were entered as "world," while individual state and local designations were entered as "place." Word cloud by Kathleen A. Povis.

natural history exhibitions and outreach. Taken together, the mission statements of participating institutions reveal an overwhelming dedication to education, research, Earth, and—most prominent—nature (Fig. 1.5). With conversations of climate change, biodiversity loss, and energy becoming more commonplace within society, this shared dedication positions institutions of natural history at the center of today's most challenging issues. With their mission to preserve and study nature and to educate about the world and humanity's place within it, such places of natural history are more relevant now than at any other time.

Conclusion

Natural history museums are indeed the stuff of legends. Ask your seven-year-old self to remember back to that first encounter with a real dinosaur, or that time you met a mummy. The dusty hallways and the smell of history are powerful mnemonic tools, and they are tools that should not be underestimated. Natural history museums are again finding their charge within society, answering a new call to return to their roots and promote humanity's sense of being and place within nature. Goode's triad of "preservation, research, and education" did not set up mutually exclusive categories. Together they can continue to flourish, and through exhibition we will engage today's seven-year-olds in natural history.

Famed naturalist Sir David Attenborough was asked how he viewed the role of museums in the future, to which he responded, "These objects...have a huge research potential in them...and you think 'Crikey' and you can...use it now for research in a way that the people that collected it never dreamed of" (Attenborough 2010).

CHAPTER TWO

Natural History and the Nature of Science

"Men love to wonder, and that is the seed of science."

—Ralph Waldo Emerson

The natural world—nature—is infinite. It is the physical universe and all that it contains, past, present, and future. The collections of museums, often stored within dark rooms along rows of endless shelving, are humanity's records of nature. And while a single object represents an event or individual, thousands of a particular type contain the story of nature and its history. As the storyteller, museums provide their visitors with limitless opportunities for discovery and inspiration.

But objects and their basic, *what-is-it* interpretation aren't enough. With the issues of biodiversity loss, climate change, and Earth stewardship departing the realm of the esoteric and becoming increasingly relevant within the decision-making processes of society, today's natural history museums find themselves in a unique position. Embedding the process and characteristics of science—the nature of science—within natural history displays can lead to heightened understanding, increased scientific literacy, and the empowerment of visitors. It is the responsibility of the natural history museum to offer its visitors not only knowledge but also the capability to understand science as a process and, in turn, equip them to better evaluate the critical issues of the current day. To this end, in order to properly promote science and scientific literacy, a basic grasp of inquiry and the nature of science must and should be understood by exhibition teams working in all science-based settings.

Nature and its History

Natural history is everything. Eight million years ago, deep within the great continent of Africa, a curious ape stood upright and looked out upon the vast, expanding grasslands. With a few short steps, it left behind the forest that it—and for millennia its ancestors—called home. A new day had dawned. The human story began. From this primitive form, a parade of descendants evolved

Sarah J. Chicone and Richard A. Kissel, "Natural History and the Nature of Science" in *Dinosaurs and Dioramas*, pp 25-33.

to produce a rich family of species. By 2.5 million years ago, technological advances had led to simple stone tools. The mastery of fire appeared around one million years ago, along with the rapid development of larger brain sizes. Nearly 200,000 years ago, our own species—*Homo sapiens*—first appeared and found success almost immediately.

The term *origin* is relative, of course. Determining a starting point for the human story is a purely subjective—and enticing—exercise. Indeed, humanity's origin is often presented as the moment our lineage stepped away from other apes eight million years ago, but equally acceptable entry points are found throughout life's much longer history. Sixty-five million years ago, for example, a six-mile-wide fireball collided with the Earth. Ecosystems collapsed around the globe, and life's course was forever altered: half of all species fell to extinction, including all of the large dinosaurs. Had the impact not occurred, their reptilian lineages would have no doubt persisted for millions of years, if not more. But the demise of the dinosaurs allowed mammals, which had already existed for more than 100 million years, to thrive. With dinosaurs no more, mammals flourished and, at that moment, gave rise to the basic forms that dominate the world today. This post-extinction Earth was now home to primitive horses, early relatives of cats and dogs, and a peculiar group of mammals that called the trees home: primates.

Traveling even further back in time, we can trace our origins through the emergence of the mammal family tree just over 200 million years ago; the evolution of tetrapods—animals with limbs and digits—from fishes 360 million years ago; the first appearance of animals with backbones more than 500 million years ago; and, around 600 million years ago, the development of the very first animals: soft-bodied worm- and frond-like creatures. Ultimately, the origins of humanity, as well as every organism living today, can be traced to the very first, bacteria-like organisms that lived on a primitive planet some four billion years ago.

Today, seven billion humans inhabit the planet, with an expected billion more by 2020. It's incredible to realize that knowledge of our earliest kin and deep ancestors is based on the delicate, fossilized fragments of few individuals. But these fragments tell our story, and it is a story firmly rooted in the disciplines of chemistry, physics, biology, geology, and—most recently—anthropology. Humanity is not separate or above beast and root but, quite contrarily, embedded within nature. We are a part of nature and its history, and the same can be said of any lineage, from tiny bacteria to towering redwoods. The history of the natural world and all of the stories that it contains represent an intertwining story of the living and physical worlds. Though the convenience of packaging scientific disciplines into separate textbooks and exhibitions tends to dominate the classification-based thought process of the human mind, these disciplines are inseparable. They inform and complement one another, and they are often best presented as such.

Nature and its history hold an untold number of stories. Those that we choose to tell within the halls of our world's museums are often presented as isolated, esoteric vignettes of this history, but the best presentations of science-based topics realize the interconnectedness of all things natural. That is, they appreciate the context. Halls within museums will forever carry different names and see their creation developed by different teams of staff and contractors, but key to the presentation of our natural world is remembering that the stories we choose to tell are unified under the umbrella that is the natural history. Natural history is everything.

Science as a Way of Knowing

What is science? Images of well-creased lab coats and Goldbergian systems of flasks and funnels spring to mind. But, at its very core, *science is simply the attempt to understand and explain the world in which we live.* These explanations are endless and vast—whether they address the origin of the universe or the inner workings of our own cells—and they are critical for humanity's sense of being and satisfaction. We are a curious species. And science is our way of knowing.

The term science is derived from the Latin term *scientia*, which translates to "knowledge." But defining science as simply a body of knowledge is an extremely limited view: it presents only the results of science. Science is so much more; it is also the method by which humanity acquires that knowledge. In that light, science is dynamic, science is creative, science is emotional, and, most significantly, science is a *process*. It is the process by which human beings ask questions about their world, seek and consider information that leads to answers, and then share this knowledge with others. It is a process of understanding that is both empirical and rigorous, and it is beautiful for the very basic, guiding principle that scientific conclusions are evidence-based and, in the light of new evidence, are tested time and again, generation after generation. As definitively stated by the late Carl Sagan, "Science is more than a body of knowledge; it's a way of thinking" (Sagan in Head 2006, 143). To present science as paragraphs of endless facts, cases of aging specimens, and tongue-twisting vocabulary terms is to perform a disservice to science and those wanting to learn about and engage with their world.

Increasingly important in any scientific instruction—within both the formal classroom and informal venues like museums—is a discussion of the *nature of science*. Commonly considered in the context of formal classroom education (e.g., Kruse 2005), conversations regarding the nature of science (NOS) are equally important within informal environments. NOS refers simply to the characteristics of science. In greater detail, as defined by Lederman and Lederman (2005, 53), NOS refers to the "characteristics of scientific knowledge that necessarily result from the scientific investigations that scientists conduct to develop knowledge." Importantly, this expanded definition stresses the process of science and not just its results. Following Lederman and Lederman (2004; 2005) and Lederman (2007), discussions involving the Nature of Science include the following points:

- *All scientific knowledge is empirically based and derived from observations of the natural world.* The conclusions of scientists are not based on whims or best guesses but, instead, must be based on data and evidence. The greater the evidence, the stronger the conclusion. And, importantly, this data is available and observable to all, allowing other scientists to reexamine conclusions using the same observations and data, and adding to it. The conclusions of science are therefore testable and those tests are repeatable. This process of science is evidence-based and rigorous. Faith, in contrast, is belief that persists in the absence of evidence. This distinction is extremely relevant for exhibition developers of science-based institutions. The presentation of creation stories within an anthropology-based exhibition should receive a different treatment than the origin and evolution of life within a paleontology- or biology-based exhibition; the latter is evidence-based, while the former is not.
- *Observation is not the same as inference.* Observations are statements about the natural world that are directly accessible to the senses, while inferences involve a process of reaching a conclusion based on observations and evidence. It is observed, for example, that a tooth has a flattened surface. It can then be inferred that such a tooth shape was likely used for grinding plants.
- *Scientific knowledge is subjective.* Scientists often claim that their study was an objective analysis, but science is home to nothing that is not subjective. Nature exists outside the human mind, but science is our interpretation of the natural world and the process by which we make that interpretation. Scientists are individuals who bring different experiences, practices, expectations, and knowledge to their work; their work is a reflection of who they are and their history. Science is, first and foremost, a human endeavor. Thus, while it is an evidence-based process, science is subjective by definition.
- *Scientific knowledge involves human imagination and creativity.* All science begins with a question. Developing investigations to find an answer to that question can involve an incredible amount of creativity.
- *Science is socially and culturally embedded.* Scientists are individuals from all corners of the world, and science affects—and is affected by—the elements and context of the culture in which it is practiced. Political structure, philosophy, religion, socioeconomics, and social fabric can all influence scientific pursuits.

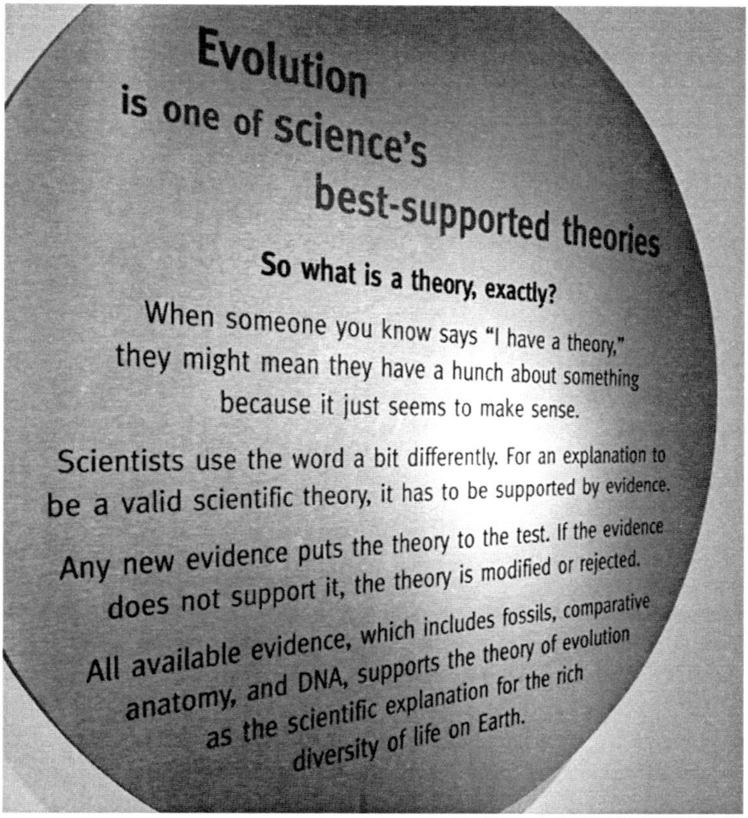

FIGURE 2.1. At the entrance to The Field Museum's *Evolving Planet* exhibition, visitors find an explanation of the term "theory," preparing them for its use in science and the 27,000-square-foot exhibition. Photo by Theresa Fiala.

- *Theories and laws, though related, have different functions.* There exists no word as abused as *theory*. In common vernacular, it is substituted for guess, but its meaning within the scientific field is quite different. And the difference is so very important. In science, a theory is an inferred explanation for observable phenomena that is supported by a large body of evidence and has been tested time and again. So, in science, nothing is "just a theory." The theory of evolution by means of natural selection, for example, is an explanation supported by an overwhelming number of observations and an extremely large body of evidence. With the opening of its permanent gallery *Evolving Planet* in 2006, Chicago's Field Museum very directly addressed the challenge of the term *theory* and visitors' misconceptions of it, providing a definition within the introductory gallery (Fig. 2.1). A law, on the other hand, is a statement or description of relationships among observable phenomena. Despite the common notion, theories never become laws; theories explain laws.

- *Scientific knowledge is tentative.* Last, it is essential to realize that science is not static, and that trait is perhaps its most important. While textbooks overflowing with charts, graphs, and other pieces of information imply otherwise, science is and will forever be an ongoing process of discovery, questions, and investigation. Compare, for example, any book on geology, anthropology, or any discipline from the 1950s to one published today; the level of understanding will be much more complete in the latter. New discoveries by new generations of scientist lead to new information that adds to, confirms, or requires a reevaluation of existing ideas. Each new discovery is a puzzle piece, and each new piece helps science paint a more complete picture of the natural world. Science changes as the evidence dictates, bringing humanity closer and closer to a more complete understanding of the natural world.

Together, these seven characteristics can provide new insight into science as not a collection of facts, but a process of discovery and explanation. An understanding of NOS allows educators and exhibition developers to address the well-documented misconceptions that students—and visitors—have regarding science (e.g., Moss 2001; Bell and Linn 2002; Bell et al. 2009). Presenting the above list within any exhibition is certainly cumbersome, but embedding these concepts within an exhibition's storyline or narrative is realistic and a reasonable expectation. Providing insight into the history of a scientific subject, for example, can provide the perfect platform for discussing NOS (Akerson et al. 2000; Clough 2006), and such an approach can easily be incorporated into exhibitions. For example, today we know dinosaurs as one of Earth's most successful type of creature, from the diversity of their forms to the sheer number of species and duration of their existence. In recent decades, science has even firmly and confidently established that birds descended from a lineage of small, meat-eating dinosaur, leading to the unavoidable conclusion that birds are a living type of dinosaur. Contrast these views with those from the mid-20th century, when dinosaurs were simply viewed as an interesting but esoteric group of reptiles that ultimately couldn't survive in their world. This historical approach presents opportunities to embed all seven of the NOS characteristics listed above, from changing science to the scientific personalities that led the way to those changes. It and other explicit approaches can lead to an increased awareness and understanding of science as a process[1] (Lederman 1999; Khishfe and Abd-El-Khalik 2002; Schwartz et al. 2004).

Inquiry and the Myth of the Scientific Method

Inseparable from the Nature of Science is *scientific inquiry.* Inquiry is central to science, and the importance of incorporating inquiry into science learning has been understood and stressed for some time (National Research Council

1996). But what is inquiry? Scientific inquiry refers to the many ways in which scientists study the natural world. It is the active exploration for evidence-based answers to a question of interest or curiosity; it *is* the process of science. Important to understanding the process of science is to know that all scientific investigations begin with a question. Why are the dinosaurs extinct? How were the pyramids constructed? These and infinite other questions spring from our imagination- and curiosity-driven sense of being.

Following the initial question, as outlined by Lederman (2012), it is important to realize that there is no single set or sequence of steps followed in all scientific investigations. Every investigation is different, which counters the idea of a single *scientific method*—a notion presented within textbooks of nearly every scientific discipline (Table 2.1). Instead, inquiry procedures are guided by the question asked, and procedures might vary depending on the scientist performing them. Regardless of this subjective trait, the conclusions of any study must be consistent with the evidence and data collected. While the avenues of scientific investigation are many, science is—above all—empirical (Padilla 2012).

TABLE 2.1. The Scientific Method, as traditionally presented within many educational venues, from classroom textbooks to museum labels, incorrectly suggests that science follows an unchanging, linear process.

Step One	Identify a Problem / Question
Step Two	Formulate a Hypothesis
Step Three	Perform Experiment / Test Hypothesis
Step Four	Gather Data
Step Five	Draw Conclusions

Introducing inquiry within a museum setting is possible at many levels. A simple but often overlooked approach is to develop label copy that asks questions; these questions invite visitors to begin an inquiry-based thought process. Opened in 2011, the 14,000-square-foot *Dinosaur Hall* at the Natural History Museum of Los Angeles County is an example of an inquiry-driven exhibition. At the entrance to the space, an introductory panel establishes that scientists have constructed a knowledge base about dinosaurs by asking questions about them and their world (Fig. 2.2). This simple panel elegantly frames science as both a body of knowledge and a process that begins with questions. Panels throughout the galleries follow this theme, with many of them posing an overlying question. *How diverse were dinosaurs? What was ancient Earth like?* The exhibition is built upon inquiry and the Nature of Science, and it was granted a 2012 *Special Distinction – Innovative Reframing of Collections* achievement from the American Alliance of Museums for this approach.

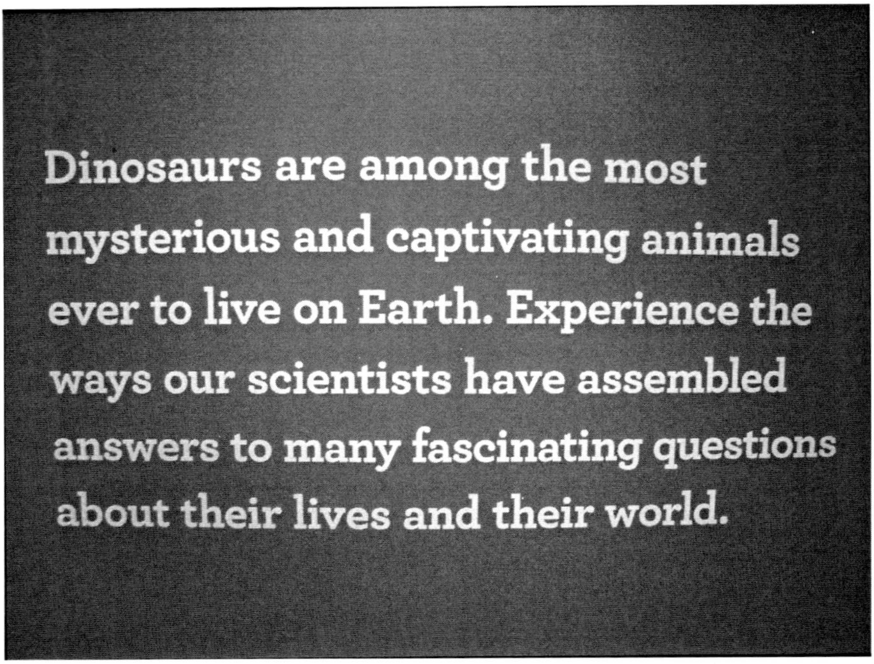

FIGURE 2.2. In only 32 words, the introductory panel to the *Dinosaur Hall* at the Natural History Museum of Los Angeles County effectively prepares visitors for their experience by introducing them to the inquiry-driven methods of science. Photo by R. Kissel.

Renowned biochemist and author Isaac Asimov once suggested that, "The most exciting phrase to hear in science, the one that heralds new discoveries, is not *eureka* but *that's funny*." Natural history museums have the power to produce such reactions, with objects as exotic as dinosaur bones, otherworldly flora from distant lands, rocks from distant planets, and fragile artifacts from ancient civilizations. The active exploration for answers to a question or a piqued curiosity—inquiry—develops a heightened understanding of science (e.g., National Research Council 1996, Fries-Gaither and Shiverdecker 2012), and any museum can provide a setting that allows such initial curiosity to bloom.

Conclusion

For thousands of years, humanity has created explanations for the natural world. But science, humanity's *evidence-based* explanation for natural phenomena, creates a singular vision that unites and directs our knowledge. The birth of modern science can be and is often traced to a very particular moment in time: 1543. That year, on the eve of those greatly constructed curiosity cabinets of Imperato, de'Medici, and others, *De Revolutionibus Orbium*

Coelestium (*On the Revolutions of the Heavenly Spheres*) by Nicolaus Copernicus and *De Humani Corporis Fabrica* (*On the Fabric of the Human Body*) by Andreas Vesalius were published. Copernicus' heliocentric model of the heavens—with Earth as not the center of existence but merely a body orbiting the sun—led to the birth of modern astronomy, while Vesalius' volume offered an unprecedented description of the human form, in both presentation and accuracy. In the nearly 500 years since Copernicus and Vesalius, generation after generation of scientists have contributed to our ever-growing understanding of our world. And each generation has been fueled by discoveries, questions, and the pursuit of these evidence-based explanations. All science begins with a question, and scientists are simply those children that never stopped asking: Why?

Providing a uniform presentation of science across museums might seem like an unobtainable goal, but it is certainly far from impossible. While gallery design and label copy will vary, science as a process of understanding will not. The average American spends less than five percent of their life in the classroom (Falk and Dierking 2010), a finding that firmly establishes the role of informal educational venues like museums as the prominent venue for public science learning. Museums possess the incredible potential—and responsibility—to educate and inform the public's perception of science. Science is not just facts and jargon; science is a process. This call to action cannot be overstated. Developers of exhibitions and educational programs within museums inherit the responsibility not only to entertain but also to help and inspire visitors to better understand not just "science" but the process by which humanity understands its origin and place within nature and its history.

Part 2

Development and Design

CHAPTER THREE

What's Gonna Work? Teamwork

"It takes two flints to make a fire."
—Louisa May Alcott

Having positioned the development of natural history museums in historical context and espoused the importance of presenting science as process, we now bring the conversation into tighter focus and begin looking more specifically at the creation of natural history exhibitions. And while subject matter and content determine the focus of an exhibition, the individuals that develop and design the exhibition are the ones that determine, create, and shape the storyline and experience for the visitor. We therefore begin our journey by focusing on this critical group of invested players: the exhibition team.

An exhibition developed by one institution will turn out much differently than one on the same topic developed by another institution, despite similar content and institutional mission statements. This phenomenon is based on such logistical factors as square footage, schedule, and budgetary constraints, but most importantly, it is based on the members of the project's team. Think about any number of natural history museums that you might have visited. Many chronicle evolution through time or the archaeological history of North America, but how is this done? Where is the focus? Which subjects, stories, or objects are highlighted, and how were these decisions made (Lynn et. al 2005)? Such decisions can be very exciting when you consider that we, as individuals within the museum profession, can significantly influence an exhibition that very often has the potential to reach thousands, and in some cases millions, of visitors.

The formalized idea of a team of contributing members designing and developing museum exhibitions has been in practice since the late 1970s.[1] Since it was first articulated as an approach, it has become the preferred way of creating exhibitions. As noted by McLean (1993, 36), the team approach replaced the vertical hierarchy of the traditional curatorial model with a "lateral and inclusive" method. The Smithsonian Office of Policy and Analysis (2002a, 14) further contends that it "represented an attempt to negotiate authority between subject matter or content experts (curators...) and subject matter interpreters (educators...)." And there is general consensus in the museum community that this system is indeed a stronger and more productive approach than the linear, trickle-down design methods of the past (Jennings 2002; Morris 2002). Isble (2010, 27) further proposes that "achieving minds-on exhibitions that engage visitors is not possible with a hierarchical or top-down approach to exhibition development. Rather, it requires teamwork and expertise from many departments within the museum." Despite all of its benefits, the museum community also recognizes that this approach is not a silver bullet, and it is also not without its own set of challenges.[2]

Even beyond the interactions among immediate team members—both internal and external—the range of various stakeholders can directly impact logistics and process. Project teams work within existing systems of institutional governance, which can be a source of either support or constraint. Stakeholders can include directors, board members, and/or connected foundations, and require good communication, chemistry, and politics.

A further consideration is the role of external funders in the design and development process. Exhibitions are often sponsored directly by private philanthropists, foundations, corporations, or government agencies. As Alexander suggests in her decades-long studies of art museum exhibitions, "There is a very clear connection between external force (funder) and organizational output (exhibition)" (Alexander 1996, 798). Alexander's work (1996) has shown that the funding environment ultimately impacts the kinds of exhibitions developed and the way they are presented. It is not just a matter of handing over a check; the involvement of external stakeholders in the actual design and planning of an exhibition may also be included in the interest of a collaborative spirit. In fact, this is increasingly becoming an important component alongside a growing "private and commercial climate" (Gray and Kendzia 2009, 173).

Nevertheless, corporate involvement in the process can result in observable consequences to both educational content and experience—what Gray and Kendzia (2009, 168-169) describe as "spectacular but largely superficial exhibitions." With an increasing trend toward private sector control of public services (ibid.), museums should "develop internal responses to funding pressures" (ibid., 174). Defining an institutional ethics policy on corporate sponsorship will help establish boundaries on content control and pressure by external stakeholders within an exhibition team.

Teamwork: The Good, the Bad, and the Ugly

McLean (1993, 40) insists that developing an exhibition is an act of collaboration, and that is the view—with an emphasis on *collaboration*—advocated here. Multiple minds and voices add dimension to both content and approach. In a perfect world, this scenario provides room for the good ideas to simmer to the surface, and the bad ideas to filter out the bottom. It brings together diverse skill sets, experiences, and expertise to ultimately create a better and more impactful product. But focus and cooperation are key.

Group work has the grand potential to be messy, unfocused, and unreliable; true vision and innovation can be stifled by *groupthink*—a "mode of thinking people engage in when they are deeply involved in a cohesive in-group, when the members striving for unanimity override their motivation to realistically appraise alternative courses of action" (Janis 1982, 9). Groupthink commonly leads to faulty decisions and poor product (Janis 1972). Teams might also find themselves handicapped by unyielding and dominating members. But careful planning and implementation can result in a stronger process—both for members of the group and for the institution—and, consequently, a stronger product, a culmination and synthesis of creativity, focus, and expertise.

So what, then, makes a "good" team? There has been much written to define the various models and approaches to team-centered design (Kamien 2002), as well as the various players needed to achieve that special *je ne sais quoi* (McLean 2001; Bine 2006; Jennings 2006; Sims 2006). But, like much of what's discussed in this book, there are many factors that impact approach, not the least of which are institutional size and mission, project scope, staff expertise, project budget, timeline, and content. It likely does not come as much of a surprise that good teams can be as varied as the projects they work on. Some museums have institutionalized the team development of exhibitions, designating project partners at various stages in the process, but for others it is a much more fluid process, dependent on staff interest and experiences as it relates to content and approach. Both strategies, when implemented well, can have positive impacts on the ultimate product and visitor experience. Faron and Curran (2007, 33) suggest that "there's no magic in team building," and it can be further extended that there is no *magic team*. But there are a number of considerations and strategies that can lead to a smooth and efficient process capable of producing great results.

The Players

Exhibition teams can include a diversity of players—internal staff, outside consultants, invested community members, and volunteers. Although by no means an exhaustive list, common internal and external players may include the following:

- *Project Manager.* The project manager is in charge of facilitation and information distribution, and is the sole individual responsible for ensuring that the project—its many tasks and milestones—stays on budget, on schedule, and on task (D. Whitemeyer, pers. com.). The Project Manager is often the face of an exhibition's development, in both formal and press-related events.

- *Exhibition Developer.* The exhibition developer often wears multiple hats, serving not only as researcher but also communicator. They offer structure to the efforts of everyone involved, and lead the development of content and approach along the way (Jennings 2006). Working closely with content specialists, they develop an exhibition's Big Idea and goals (see Chapter Five), and they are often responsible for writing label copy, while also finding and providing an exhibition's voice and tone. Throughout the process, exhibition developers ensure that content, text, imagery, and design are complementary, combining to tell a story that the visitor will find both engaging and accessible.
- *Content Specialists/Curators* are the experts in the field—the "knowledge keepers"—and they ensure that all content is both accurate and current. Content specialists work closely with *every* aspect of a project. They are the content resource for developers, designers, illustrators, model makers, and animators, and they are ultimately responsible for the correct presentation of content in the exhibition.
- *Exhibition Designer.* Exhibition designers lead the process of visual thinking, both graphic and 3D design, and then determine a way to make the vision a reality (Sims 2006). Designers work closely with developers to ensure that the visual presentation of an exhibition matches the tone of the labels and supports the exhibition's messages.
- *Collections Personnel/Registrars* help recommend, facilitate, manage and care for exhibition objects. They work closely with developers and content specialists to select objects at a project's onset, and then work with production specialists for determining proper conditions for display, such as light levels, humidity concerns, and mount making.
- *Educators.* Once an exhibition opens, it is often ceremoniously passed from the exhibition team to an institution's educators, who develop activities, tours, and other programs centering on an exhibition's content. It is advisable that educators participate in team meetings, so they can contribute to the pedagogical aspects of the exhibition and are well informed about the exhibition's content.
- *Evaluator.* Evaluators focus on visitor experience and conduct "scientifically defensible studies for the purpose of testing-improving-documenting the use and effectiveness of museum experiences" (Hayward 2006, 79). (See also Chapters Four and Nine for discussions of evaluation.)
- *Exhibition Fabricator.* Fabricators, both internal and external, produce an exhibition's physical presence, including both 2D and 3D elements—ranging from general cabinetry, graphic panels, and labels to enviroscapes and immersive environments.

- *IT Staff.* An institution's Information Technology (or IT) staff have a good understanding of the capacity of an institution to support and maintain technology. It would not be wise to include technological elements in an exhibition that no one on staff knows how to work, or how to maintain.
- *Media Designers* use technology, or multimedia, to tell a particular story. (See also Chapter Six on interactive elements and engagement.)
- *Visitor Services Managers.* These internal staff members keep visitors' needs in focus; they often have intimate knowledge of the museum floor, including traffic flow, staffing requirements, and capacities.
- *Marketing Specialist.* As the exhibition development process nears completion, the entrance of marketing staff can ensure that the exhibition's marketing matches the content, visual design, and tone of the exhibition.

No matter who comes together to create your team, the key element of a successful project is process, and that process must be supported by a strong, focused Project Manager. Whether the Project Manager only manages the project or wears many hats, there is a significant need to ensure that the project remains on task and target regarding schedule and budget, and—just as important—maintains a healthy team dynamic. It is the Project Manager who's responsible for ensuring a successful process (Bine 2006). Even within a collaborative framework, strong leadership of process is needed and real decisions need to be made in real time; the Project Manager moves this process forward.

In or Out: Internal Staff

Exhibition teams can be composed entirely of staff members within a particular institution. Pooling the talents of existing staff is a great way to increase investment, understanding, and ownership of institutional outreach. Exhibitions tell the stories of Earth and its peoples; they serve the community, they serve the science, and they contribute to the construction of social knowledge and our understanding of the world in real and significant ways, and whether you reach 100 or 1,000,000, this is exciting! Exhibition projects are undeniably *a lot* of work, but they can also be *a lot* of fun, bringing together creativity, content, and visitors in the fulfillment of institutional mission and purpose.

Another benefit of using in-house staff is that such a strategy can ensure an intact and cohesive team throughout all stages—from the initial brainstorm to the exhibition's opening. By being involved in the entire process, from beginning to end, a core group of invested team members can help maintain a consistent message. If this scenario is unrealistic, especially for small institutions where staff members often have responsibilities that intersect departments, another viable scenario is to incorporate players on an as-needed basis as the project phases wax and wane. This strategy can supplement group expertise when needed, as needed, to contribute to development and design.

In or Out: External Partnerships

When considering external partnerships, a team must answer an important question: What are the strengths and weaknesses of the existing exhibition team? By understanding your team's abilities, as well as the realistic capacity of your institution, you can anticipate where the project would benefit from external assistance and expertise. A museum with a smaller exhibitions shop that is designed primarily for maintenance and not fabrication, for example, might consider contacting a local cabinetmaker or fabricator to help build some, or all, of the planned 3D display elements. Larger institutions often benefit from the background and experience of exhibition design and development firms that have familiarity realizing complex and comprehensive exhibition projects, within budget and on schedule. And any given project might include a considerable technological component that might fall outside the scope of staff expertise. Independent web and technology developers can assist with interactive elements, online websites or apps, and other technological aspects that a team would like to include in the project design. Independent consultants and contractors can add valuable insight and experience to projects regardless of their size and scope.

Just as members of an in-house team should be recognized and respected for their unique skills, the utility of outside practitioners in the field should never be underestimated. Whether it's seeking advice through a relevant listserv or a regional consortium of institutions of similar size and mission, partnerships between your institution and others can serve to increase your capacity by combining diverse and complementary skill sets. While it can be overwhelming for both small and large institutions to navigate the complex world of museum consultants, there are a number of resources to help with this process. Web-based resources linked to professional and social groups, as well as regional association listservs, are valuable tools for posting questions, getting answers, and sharing information.

One of the most daunting aspects of working with contractors is choosing the best match for your particular project. Regardless of the exhibition's scope, it is advisable to start with a Request for Proposals (RFP).[3] Consider the exhibition elements you have in mind; where do you need additional help? It might be the design, the fabrication, the multi-media components, or maybe it's the entire project. It stands to reason that if you know what you need, you will know what to look for. Having someone with experience writing strong RFPs and understanding your own institution's purchasing protocols also helps. And remember, contractors are working *with you* and *for you.* They are not just your employees but also your collaborators—they bring significant skills to the table that fill the needs you have identified. All team members' goals should be convergent.

In addition to contracting agencies that might assist with the design or fabrication of exhibition elements, the following external consultants can also provide a great deal of assistance to an exhibition team:

- *Indigenous Communities.* Museums serve wide and diverse publics, including obvious constituents of volunteers and visitors, as well as—in the case of many natural history museums—indigenous and descendant communities. In addition to internal staff and external consultants, project teams can also include other invested members from the variety of communities they serve. The Indian Advisory Committee (IAC) at the Natural History Museum of Utah serves in just such a capacity. The museum was recently re-envisioned with a new facility, the Rio Tinto Center, which opened in November 2011 (Sampson and George 2004). According to the museum, the IAC assists in "planning programs and exhibits that meet the needs of Indian communities in the Intermountain West, and in implementing NAGPRA, the *Native American Graves Protection and Repatriation Act.* The Indian Advisory Committee has been integral in the interpretation of Utah's Native Peoples in [the] Museum" (http://nhmu.utah.edu/museum/about/iac).

- *Volunteers.* Volunteers can also bring significant experience and insight into the exhibition design process. Within many institutions, the educational activities are led not by staff but by dedicated volunteers, so they often spend valuable time on the floor watching, observing, and interacting with visitors. Draw on their unique relationship with visitors, and on their often-passionate relationship with the institution and its mission. The key is to listen to their voice; let them be heard in the process. Museum staff often have a larger responsibility to the institution and the profession, but volunteers provide an incredible wealth of information, provided that they are heard.

- *Visitors.* There are numerous ways to engage the larger community in the process of exhibition development. Nina Simon, Executive Director of the Museum of Art and History in Santa Cruz, discusses the nuances of this approach in her book *The Participatory Museum* (2010), which presents techniques for institutions to invite visitor participation while promoting institutional goals. She describes these relationships as falling under two categories—those that help guide (consultative) and those that help create (co-development).

  Consultative approaches seek community member advice and guidance on a particular project or one of its aspects. Focus groups, round tables, and prototype parties are designed to engage end users with the design process (Simon 2010). Activities that involve the community need to be focused and meaningful for those targeted, as well as the institution. Simon suggests that a significant outcome of the collaborative process is the way it impacts staff perceptions of visitors and community members—changes happen when these groups are viewed as collaborators not mere consumers of an end product. According to Simon (2010, para 82), participants "gain new respect for their [visitor] interests and desires." In consultative approaches, visitors are generally not engaged in the creative aspects of design, but instead in what LaBar (2009, 36) describes as a "market, or audience."

Consultative approaches can target specific groups by age, gender, ethnicity, and expertise. Many natural history museums have access to significant experts, curators, researchers, and leading thinkers in their field. Relationships between exhibition developers, designers, and scientific researchers can be exciting and incredibly fruitful. There is hardly anything comparable to the passion of committed scientists; translating this passion and making research accessible to general audiences is appropriate and doable.

Compared to a consultative approach, there is significantly more at stake when a team co-develops an exhibition with community members. *Co-development* introduces a discussion of institutional authority, and this discussion is potentially uncomfortable. Places of informal learning that promote science—natural history museums, zoos, aquaria, botanical gardens, nature centers, science centers, etc.—possess an interesting relationship with authority. As Simon (2006, 71-72) states, they "do not value multiple perspectives on basic interpretations of science"; in fact, "established scientific theory is not open to visitor reinterpretation." So where does this leave conversations on co-development, institutional authority, and participatory culture that are happening in the larger museum field? Stein (2012, 218) suggests that it comes down to "an attitudinal shift among museum staff concerning an approach to authority as it relates to participatory culture." He advocates a shift from authoritarian, which requires "the museum to assert its authority by definition," to what he describes as authoritative, which "demonstrates the expertise of the museum's staff in a participatory context that recognizes the museum's role in facilitating access to cultural objects and information.... [in this case] authority is ascribed to the museum by the audience, and not the other way around... the audience is the central determiner of value, but the museum is responsible for building an appreciation for cultural access and preservation" (Stein 2012, 218).

While still minding the absolute necessity of presenting scientific theory correctly, natural history museums have been experimenting with co-development for years, as they have sought to engage indigenous and descendant communities, as mentioned above. Such institutions have just begun to navigate shared curatorial practice, and there is much still to be done (Rand 2009). The inclusion of colonized peoples in natural history exhibitions has been significantly problematized, for, in part, what anthropologists MacDougall and Carlson (2009, 167) suggest was the institution's secondary purpose of reinforcing the "power and authority of colonial regimes." Natural history museums cannot escape their past, but by properly preparing for their future, they can work toward repairing old relationships and maintaining new ones.

Shared curatorial practice can be a risk for both institution and native groups, but the rewards can be significant. In fact, the rise of tribal museums has occurred in response to what Anthropologist Susan Sleeper (2009, 4) suggests is a global "demand to create alternative narratives and give force to formerly colonized people." There is no reason that without much reflection and engagement, natural history museums, with their mission to present the history of Earth and its peoples, cannot also fill that role.[4] There is a meaningful difference between incorporating what Rand (2009, 130) calls a "benign native voice" into exhibitions and creating space for the native control of content—creating "project equals" versus implementing a "patronizing spirit of inclusion." Natural history museums large and small have significant artifact collections that native groups can use as "contact zones," which can begin to bridge the "wide gulf between native experience and memory and the superficial knowledge of a non-native audience" (ibid.).

Ultimately, assembling a successful exhibition team is a complicated enterprise that deserves considerable time and attention, but tested and successful models exist. Consider the needs of the particular exhibition, the invested players—internal and external—and how they might come together to actualize the collaborative process. Whether it is a group of three fulltime staff or a cadre of scientists, consultants, volunteers, and other invested communities, the cast of players directly impacts the end product. Who does the saying is as important as what you want to say and how you want to say it.

The Process

As we have seen, teams can come in a variety of shapes and sizes, with varying degrees of skill, expertise, and investment. But just as important as assembling the right team (internal, external, community, or volunteers), of course, is devising the right process. A variety of processes are possible and inevitable, but keeping commitment, the accessibility of information, and communication at the forefront of development will help ensure a truly collaborative framework. As such, these factors should also structure the design and development.

It is important to outline communication strategies from the very beginning. When a project launches, the excitement embedded in the possibilities is often palpable. People are excited, energetic, optimistic, and driven; the idea is to capitalize on and maintain this momentum throughout the entire process as deadlines become real, ideas get 'value engineered,' and the inevitable compromises occur. Clear and consistent communication strategies are needed to keep the process moving forward. This type of planning helps ensure that expectations are met—teammates know what to anticipate from not only the process, but also from each other. For internal staff, these expectations are often implicit; making them explicit adds transparency and allows teams to anticipate potential points of conflict before they happen. When collaborating with outside consultants, incorporate expectations into the deliverable schedule and contract—*what* is expected and *when* should be clearly outlined, with payment hinging on successful completion of deliverables, or agreed-upon milestones within that completion schedule.

There exists any number of resources to help teams consider some of these larger logistical issues before they jump into the process. We recommend thoroughly examining the following key areas:

- *Project roles and responsibilities: Who will do what and when?* Thinking through the deliverables and schedule at the beginning of a project clarifies expectations between and among group members. The production of Gantt charts or Critical Path Method (CPM) schedules greatly assists this need.
- *Communication: How will the team communicate during the project?* Do you plan to schedule monthly or weekly meetings, and will there be online collaborative tools, such as wikis, Google docs, or shared resources on an institutional server? Is there a need to copy certain players on all e-mails related to the project? Will there be a single point of contact for all external consultants? And how will questions from team members be answered? The development of a Project Management Plan (PMP) at the beginning of the project helps clarify and formalize the communication process, while regularly written status reports help keep things running fairly smoothly. Team members will feel more invested if they realize that their contributions are being valued, considered, and responded to.
- *Collaboration: How will the team ensure that the process is collaborative?* This aspect calls back to the first point of project roles and responsibilities. Compromise is a key aspect of collaboration and is built on trust and patience. Remember, everyone should be working for the same end goal.
- *Deadlines and scheduling: How will the submission of individual pieces be handled?* Be sure to communicate any known problems or problematic dates or times that need to be considered. Also be sure to allow the group

enough time for appropriate vetting and feedback. Different deliverables will necessarily require different degrees of engagement from the group. The schedule should consider and reflect this asymmetry; exhibition development is more akin to a drunken stupor—with starts, stops, and direction changes—than a straight beeline to an end game. Make room for the evolution of ideas, but also keep in mind a project's narrowing flexibility as it moves forward; as described by Chiodo (2010, 22), the "ability to alter [the] design of a project diminishes over time."

- *Draft submittals and final products: How will the review and approval process for project deliverables be handled?* Who will review text, illustrations, and graphics, and at what stages? These questions are very important, especially in the context of a science-based exhibition. Even the slightest tweak in text can change the meaning of a scientific statement. According to a 2012 report by the American Alliance of Museums, museums are trusted authorities, with 77 percent of adult museum-goers ranking museums as "equal or higher in trustworthiness" than any other source of information. Embrace this respect and expectation. Content authorities should be involved with the review and approval of every aspect of an exhibition. By knowing the project timeline and considering reviewers' schedules upfront, team members can help to build a comfortable process and, again, ensure clear expectations.

Beyond just discussing the preceding issues, a team should physically document their decisions in the form of an informal group contract. Otherwise, a team can end up with the scenario described by Braden (2010, 6), where everyone is "nodding in agreement but all picturing something different." This attention to detail regarding communication and logistics extends to all aspects of the project, and it is better to hammer out these details when the project is still fresh and the process new.

Conflict

Time has taught us that the "best-laid plans of mice and men oft[en] go astray." Even projects with institutionalized teams and roles can encounter the unexpected. The topic of managing conflict within exhibition teams has been covered extensively in the literature (Belew et al. 2010; Chiodo 2010; Cohen-Stratyner 2010; Isble 2010; Rabineau 2010; Braden 2010). As suggested by McLean (1993), teams can suffer from an individual's need to stand out played out against his or her need to belong.

Creating exhibitions is not about the individual. It is about the desire to teach and share a story of our world with thousands—even millions—of visitors, expanding their knowledge about the universe in which we all live. To be such an influential educator is a rare and unique gift. Developing exhibitions is a great responsibility. Be humbled. The process deserves respect. And those that participate in the process should recognize that respect and work with each other to create the best product possible. To be consumed by disagreements during exhibition development is not in the spirit of any museum's mission and goals. Bringing knowledge, experience, and a point of view to any team is critical, yes, but bringing a positive personality and attitude is just as important. What works for your group will depend on the personalities and the players, and we recommend following the advice of Braden (2010, 6) to "accept friction, embrace it, but be prepared."

Conclusion

Building your exhibition team involves not only assembling the appropriate cast of professionals but also managing their roles, responsibilities, and personalities. Knowing a group's strengths and weaknesses is of great value; anticipating needs keeps deadlines tight and achievable. Diversity of content, audiences, and institutional missions necessitate a range of approaches, but whether you have a small team of internal staff or a large team of many and diverse players, a commitment to collaboration, communication, trust, and respect is essential to the process.

CHAPTER FOUR

Knowing the Audience

"Wisdom begins in wonder."

—Socrates

Every year, more than 850 million visitors pass through the doors of American museums (American Alliance of Museums 2012). That's more than 70 million per month, more than two million per day. Museums and their influence are far reaching, both in numbers and audience. Visitors from all socioeconomic and educational backgrounds explore museums, and the number of museums continues to increase at an accelerating rate (Lord and Lord 2001). While a museum's collection serves the scientist, the exhibitions serve the visitor, and those fortunate to participate within the exhibition development process are also serving—above all else—the visitor. But who is the museum visitor? And how do they learn? Each visitor brings into every exhibition a suite of interests, prior knowledge, experiences, and expectations. And those characteristics vary not only among visitors but within the same visitor, depending on unpredictable social circumstances. A consideration of audience is important to the development of any successful exhibition, as that consideration and classification of visitors informs and impacts the development and design of exhibitions.

"The Visitor"

While many within the museum profession continue to view exhibitions as primarily a method of education (Lord and Lord 2001), this approach represents a limited vision and understanding of exhibitions' role within society. Educating visitors on the world in which they live, and the science by which we know it, is certainly a goal of exhibitions, but it is not the only goal. With many in modern society viewing their leisure time as an opportunity to expand their understanding of themselves and their world, the perception of museums as the intersection of learning and leisure is increasingly important, and satisfying a series of identity-related visitor motivations will lead to increased visitor satisfaction.

Visitors are walking into museums to meet *their* goals, and Weil (1999) emphasized museums' need to transform themselves from being *about something* to being *for somebody*. To this end, Falk (2006, 2008, 2009) provides a framework that classifies visitors based on, not their actions, but their motivation-driven needs, establishing the following five categories of museum visitors:

- *Explorers* are those visitors driven by curiosity and possess a *general* interest in the content presented within museums. They view themselves as curious people who enjoy learning new things, and they view and value the museum as a place to learn and expand their general knowledge.
- *Experience seekers* regard the museum as an important destination. Visiting a culturally significant institution that displays iconic objects motivates these visitors, who are often tourists. Those visiting Chicago's Field Museum to view the *T. rex* Sue, for example, or those wanting to see the Hope Diamond at the Smithsonian are visiting the museum to collect these experiences.
- *Facilitators'* primary goal is to satisfy the needs of others in their accompanying social group. Parents taking their children to the museum, or a local visitor bringing out-of-town guests, are examples of facilitators.
- *Professional/hobbyists* comprise the smallest category of visitors. They bring strong content knowledge and visit to satisfy *specific* content related objectives. A birdwatcher, for example, might seek out a natural history museum's ornithology halls to see and learn more about the birds of a specific region. Museum professionals and photographers might also be members of this category.
- *Rechargers* (formerly termed *spiritual pilgrims*) are seeking a contemplative experience; the motivation for the visit is to reflect, rejuvenate, or surrender to the beauty of a space. Museums in their eyes are restorative places that allow them to escape the outside world and have time with their own mind.

A simplified model of visitor interaction with exhibitions is the often-cited (e.g., Parman and Flowers 2008) three-category system of *streakers, strollers,* and *studiers,* based on Perin's (1992) *streakers, strollers,* and *readers.* As defined by Monroe et al. (2006):

- *Streakers* move quickly through exhibitions, scanning for points of particular interest, but rarely lingering for long. Since they pay little attention to details, they may form broad impressions or take in bold messages, or they may traverse an exhibition without being affected at all.
- *Strollers* move more slowly, paying more attention or less at various places. They are exposed to many more basic messages, and they may pick up details here and there.

- *Studiers* are conscientious and diligent exhibition visitors who move very slowly through a gallery, trying everything and reading all of the text. Sometimes referred to as students or scholars, studiers often linger at single exhibitions for long periods of time.

Critical to this categorization is the amount of time that visitors interact with displays, which is used to measure visitor engagement—as defined by knowledge acquired. More simply, the streaker-stroller-studier model is based only on visitors' physical actions within a space, not motivation. This distinction is important. While overlap might exist, it is neither required nor predictable. One can easily see the professional/hobbyist equated with studiers, for example, but a recharger can assume all three roles. A recharger behaving like a streaker, and not acquiring much new knowledge, still fulfills their need by occupying their needed space, resulting in a satisfying visitor experience and, by extension, a successful exhibition. As the exhibition development process evolves from topic selection and team creation to opening day, consideration of visitor typology and motivations will ensure a more rewarding final product.

Learning in Informal Environments

Many if not most visitors *expect* to learn during their museum visit, and many authors have extensively explored the subject of visitor experience and learning in museums (e.g., Hein and Alexander 1998; Falk and Dierking 2000; Hein 2006; Lindauer 2007; Kelly 2007). Considered most broadly, visitors to museums engage in free-choice learning. Free-choice learning is self-directed, voluntary, guided by an individual's interests and needs, and lifelong. It is spontaneous and intrinsically motivated (Csikszentmihalyi and Hermanson 1994; Falk and Dierking 2002; Falk et al. 2006; Diamond et al. 2009); it is personalized and individualized. Museum visitors are not passive recipients of information; they actively interact with displays to make *personal* meaning of the museum experience (Silverman 1999, 2010). They control what to learn, when to learn, where to learn, and with whom to learn, depending on their identity-related motivations.

Hein (1998) provides a summary of research conducted on informal learning within museums, concluding that:

- People learn in museums; they understand and take home specific messages and ideas, and they also find moments to reflect and learn about themselves.
- Providing for visitors' practical needs—amenities, comfort, way-finding information—increases learning.
- Visitors incorporate messages and information found within exhibitions into their personal experiences and understanding of the world.

- Museum exhibitions are effective when developed as informal, free-choice environments, which carry different intentions than formal classroom education.
- Visitors are unique and make unique connections within museums.
- Wonder, exploration, and discovery are valued educational experiences within exhibitions.
- Museum exhibitions must consider visitors' prior knowledge when presenting information.

It is important to remember that learning is a cumulative, lifelong process of making meaning and finding connections (Roschelle 1995; Falk et al. 2006). As a result, many factors contribute to and influence learning in museums. Every visitor is different, with diverse educational backgrounds, life experiences, interests, and social constructs; these qualities directly influence learning and a visitor's time within a museum. Informing learning is contextual, and Falk and Dierking (2000) provide a cohesive model of visitor learning, identifying eight factors—found within three distinct contexts—that significantly contribute to the quality of a museum experience:

- Personal Context
  - A visitor's motivation and expectations
  - A visitor's prior knowledge, interests, and beliefs
  - A visitor's ability to control their learning
- Sociocultural Context
  - A visitor's interaction with members of his or her own social group influences learning.
  - A visitor's interaction with individuals not associated with his or her own group, such as museum docents, influences learning.
- Physical Context
  - Within an exhibition, advance organizers and orientation devices, such as maps and headers, provide a sense of security within a new environment, fostering learning.
  - An exhibition's design must help the visitor prioritize, organize, and correctly consume information.
  - An exhibition should reinforce events and experiences outside the museum.

Individually and collectively, these variables influence museum-based learning, and the absence of any lessens the visitor experience to varying degrees. Falk and Dierking (2000) termed this framework the *Contextual Model of Learning*, and it is based on the idea that learning is the integration and interaction of these three contexts.

A further consideration of learning within museums is the distinction between *affective* and *cognitive* experiences (Lord and Lord 2001). The more traditional view of museum learning revolves around cognition—learning and knowing facts about objects or a topic or, in other words, the accumulation of knowledge. But a more prevalent visitor experience within museums is that of the affective. Affective experiences do not involve the learning of facts but instead affect visitors' attitudes and values about their world. They emphasize the value of wonder, exploration, and the reward of expanding the mind (Hein 1998, Alexander and Alexander 2008), providing a transformative experience as visitors find personal meaning in displays and objects. Both take-away experiences are signs of a successful exhibition.

Science in Informal Environments

A Gallup poll published in June 2012 showed that 46 percent of Americans believe that "God created human beings pretty much in their present form at one time within the last 10,000 years or so." These results reinforce a consistent and considerable rejection within the United States of evolution as the explanation for life's diversity; the overwhelming scientific consensus is that the hominin[1] family first evolved some 8 million years ago, with our species—*Homo sapiens*—appearing around 200,000 years ago. Science is a complex and far-reaching field, comprised of many disciplines that are each continually revised as new discoveries refine known details or lead to totally new ways of thinking. Expecting every citizen to hold a firm grasp of the sheer volume of knowledge that is "science" is not reasonable, and established worldviews may present barriers to science acceptance (e.g., Mazur 2010).

With these challenges recognized and firmly in place, along with considerations of how people learn in museums, what aspects of museums position them at the forefront of increasing scientific literacy among an accepting population? First and foremost, museums are trusted authorities; they are places where visitors expect to find reliable, authentic, and comprehensible presentations of objects and ideas (Pitman 1999). It is a trust that not only is earned, but also should be respected by any exhibition team. Though the members of any team are all valued and deserving of respect, content is king at the end of the day. No model, illustration, or design element should see production without first receiving approval from the exhibition's content specialist. Every label should be thoroughly researched during writing, with content references noted for future use or need. Review of label copy should also include as many content experts as manageable, with at least two rounds of review built into the schedule. These reviews will ensure that the content references are up-to-date and reliable, and that the final wording of the labels is correct. Within science-based labels, even a single word or sentence structure can create or imply an incorrect statement. "It's not correct, but nobody will know," is never an acceptable approach; it is that very reason—that the visitor would never know—that the content *must* be correct.

Secondly, museums provide the opportunity to present the evidence of science—the objects. Reconstructions of ancient civilizations, or primeval ecosystems, are based on the objects we find, no matter how fractured or fragmented the artifact or specimen. Objects are active, not passive, elements of any exhibition. They are critical to the story, and their selection should be viewed as such. Each object is a piece of a puzzle; together, along with the labels and design elements, they complete the picture to provide the visitor with a clear narrative. Embrace specimens and artifacts as not only objects of beauty but also the raw material of science. As Reid and Naylor suggest (2005, 362), "Science must be grounded in the real, in the measurable, and in the object." Objects are the backbone of interpretation across many scientific disciplines. And they are more than just props that illustrate a narrative; they are active in the creation of the substance of the narrative. Exhibitions are not only a reflection of a discipline's preexisting knowledge but can be used in knowledge creation (Shelton 2006; Moser 2010; Rowlands 2006). Consider the role of the visitor as not only a passive recipient of a constructed narrative but as an active agent in the construction of their own meaning. It is the object that makes the museum unique. Museums, zoos, and other institutions of natural history are special because of those objects and the meaning making that stems from them.

Visitors value the authentic. Objects help visitors create meaning within an exhibition (Kelly 2007), and they trust that meaning only if they have confidence in the authenticity of the objects from which it is born (Lord and Lord 2001). With any exhibition, then, it is recommended that the actual object, not a replica, be used when possible (see Chapter Nine). The size, shape, texture, and color of a replica might perfectly duplicate the original specimen or artifact, but the indescribable sensation of seeing and possibly touching a 150-million-year-old dinosaur bone, or a hundreds of thousands–year-old hand ax creates a vastly more personal and rewarding visitor experience. Whenever possible, display the real. It will provide a heightened visitor experience with deeper meaning, and it offers the visitor a firsthand experience with an object that supports the science about which they are learning.

Know Your Audience

Visitors' engagement with an exhibition—regardless of their motivation—is the most important measure of that exhibition's success. Exhibitions ideally encourage and support inquiry, curiosity, conversation, learning, and reflection. And while the many talents and degrees that serve an exhibition team can often correctly determine exhibition themes, topics, and delivery mechanisms, asking the visitor to inform the exhibition throughout its development will only lead to a better product. In order to create more effective informal learning experiences, exhibition teams must incorporate visitor-centered approaches to planning. Enter the evaluation.

Evaluation is the appraisal and study of something to determine its feasibility and effectiveness (McLean 1993). For exhibition development, four types of visitor-oriented evaluations are commonly conducted. *Front-end studies* are performed prior to, or at the beginning of, exhibition development, and focus on visitors' interests, prior knowledge, questions, and concerns regarding a topic. During the development and design phases, *formative evaluation* tests the efficacy of concepts and approaches, from label copy to interactive elements (see the discussion on prototyping in Chapter Six). Once an exhibition is complete, *summative evaluation* examines its effectiveness, documenting how well the displays met the team's intended goals and objectives for visitor learning and experience.

The diversity of visitor types and experiences within a museum presents a host of challenges to those developing evaluations. Diamond et al. (2009) provide a series of metrics for measuring learning within museums, from basic knowledge retention to how visitors understand and think about knowledge. While basic interview-style evaluation serves front-end, formative, and summative studies, another useful tool for summative evaluations is a tracking-and-timing study, which documents a visitor's path through an exhibition, with notation on time spent at individual displays. A more thorough discussion of tracking and timing studies is found in Chapter Nine.

Creating effective exhibitions demands that the presentation of content is based upon a clear understanding of the museum's visitors. Visitor-oriented evaluation provides the exhibition team with guidance that will inform decisions throughout the development process. Finally, with any evaluation, data is key. Large sample sizes ensure that the results are as objective as possible, making difficult decisions regarding editing or retaining exhibition elements more informed.

Conclusion

People learn in museums (Hein and Alexander 1998). Museums are leisure-based environments that encourage informal, free-choice learning among their visitors. Free-choice learning is motivated by visitors' desires to enhance understanding and satisfy their curiosities about the world in which they live (Falk et al. 2006). But every visitor is unique; there is no such thing as a typical museum visitor. The population that visits museums is a diverse one, with each visitor possessing unique interests, motivations, desires, and needs. Even the same individual carries with them different interests on different visits, or within different exhibition galleries during a single day. In addition to developing exhibitions based on visitors' identity-related motivations, visitor-oriented evaluation further informs an exhibition team about their visitors, providing guidance on content development and design. Visitors matter, and understanding them is key to the exhibition development process.

CHAPTER FIVE

Finding Focus: Ideas, Goals, and Bubbles

"Look deep into nature, and then you will understand everything better."

—Albert Einstein

Every topic is potentially bottomless. From those as broad as the history of life on Earth, which spans four billion years, to those that appear more narrowly defined, such as the domestication of dogs or the cave paintings of Lascaux, the topics of science-based exhibitions bring with them a tremendous history and an incredible amount of information. For a content developer, they are an embarrassment of riches. But therein lies the potential pitfall. When presenting a topic to our visitors, when is enough, *enough*? Too little information and the visitor might leave unsatisfied. Too much and any subject can quickly become overwhelming. The key to an exhibition's success is finding the perfect depth of content presentation. The development of a Big Idea, learning and experience goals, and a content schematic—a bubble diagram—can ensure that an exhibition team stays focused and doesn't stray beyond an agreed-upon scope.

The Big Idea

Critical to the creation of any successful exhibition is the development of a Big Idea. Developing the Big Idea should occur early within the process, immediately following initial content research and conversations. The Big Idea answers the question, What is the exhibition about? In other words, the Big Idea is *the* central concept that you'd like visitors to take away after experiencing the exhibition; it is a simple, workable statement that provides the overarching theme of the exhibition, unifying individual displays, objects, and panels. Following Serrell (1996, 2), a well-crafted Big Idea "provides an unambiguous focus for the exhibit team throughout the exhibit development process by clearly stating in one noncompound sentence the scope and purpose of the exhibition." A Big Idea is a single sentence that possesses a subject, action, and consequence.

As an example, let's examine the topic of oceans. The subject of *oceans* is extremely broad, and any exhibition developed on the topic could quickly spiral out of control, creating a very unfocused product that has no cohesive theme or narrative. In order to provide proper direction, a Big Idea is needed. In the example dealing with oceans, one possible Big Idea is: *The oceans are home to a wondrous diversity of life—life that has a four-billion-year history of evolution and extinction.* Another possible Big Idea could be: *The oceans are home to a wondrous diversity of life—life that is threatened by human activity today.* In the first stated idea, the exhibition's focus is not only on today's ocean creatures, but it would also provide the evolutionary history of those creatures and their ecosystem. For the second, the exhibition would celebrate today's creatures but focus instead, in large part, on their fragility and loss due to human activity. Indeed, each idea might individually lead to an exhibition featuring both fossils—showing the history—and discussions of human impact today, but the emphasis would be very different depending on which Big Idea was chosen. The first idea would lead to an exhibition with more fossils than discussion of human impact; the second, vice versa.

It is important to keep in mind that a Big Idea is not a topic, nor is it a phrase that presents a promised intent; instead, it is a concrete statement relating to an exhibition's content and story. Examples of ineffective Big Ideas are *To show you many things about the ocean* or, simply, *oceans*. Nicks' (2001, 357) "core idea" is similar to but differs from a Big Idea in this very manner; as presented, it represents the intent of the exhibition team (e.g., "to provide a new way for visitors to understand the scientific method") but not the one main message that the visitor is intended to carry home with them. It is not the confident, concrete statement of content that is a Big Idea. It therefore lacks the focus that a well-crafted Big Idea offers.

Examples of Big Ideas used in actual exhibitions include, from the *Hall of Biodiversity* at the American Museum of Natural History, *The diversity of life is necessary for the survival of all living things; diversity is under threat primarily by human beings living in the environment, but things can be done and are being done to help.* As a compound sentence, this Big Idea does indeed break from Serrell's preceding definition, but it does very clearly present the focus and purpose of the exhibition. For the *Evolving Planet* exhibition at The Field Museum, the Big Idea carried throughout development was *All living things—past and present—are connected by, and the result of, evolution.* This one statement covers the concept presented in 27,000 square feet of exhibition space and content. No exhibition is too small or too big for a Big Idea. And no exhibition should be without one.

You may have also heard the Big Idea referred to as the main message. Rand's (1993, 146) "main message" serves the same role as, and is essentially synonymous with, the concept of a Big Idea. When crafting a main message or Big Idea, "Ask yourself what the single most important idea that you want people to leave with is. Write it down in a single sentence" (ibid., 146). Always remember the need for focus, clarity, and simplicity. Would your dentist or grandmother be able to walk away with the Big Idea after visiting your exhibition? An example of a longer, more convoluted Big Idea, from the Monterey Bay Aquarium, is *In the 1940s, Cannery Row residents John Steinbeck and Ed 'Doc' Ricketts launched a scientific expedition south to the Sea of Cortez. Fifty years later, Monterey Bay Aquarium biologists retraced their steps.* Notice that this Big Idea is a bit cumbersome; multiple ideas are expressed across two sentences, and informal evaluation by the Aquarium suggested that this message was too long and complex for visitors. A simple Big Idea is an effective Big Idea.

But just because a Big Idea is focused doesn't, in and of itself, make it relevant. Also important is that any Big Idea should relate to the institution's mission, which should always represent the foundation upon which your exhibition is built. The Museum of the Earth in Ithaca, New York, is an 18,000-square-foot facility that tells the story of Earth's history. In 2012, the temporary exhibition *Whales: From the Depths of the National Geographic Collection* was displayed within the Museum's temporary gallery. As initially developed and delivered, the exhibition consisted of 32 framed prints, labels, and two text panels; the exhibition was primarily a photography display of modern whales. To have the exhibition more clearly support the Museum's mission—increasing and disseminating knowledge of the history of life on Earth—the exhibition was expanded with specimens from Museum of the Earth's collections. Driving this additional development was the following Big Idea: *The diversity of whales seen today is a result of 50 million years of evolution and, more recently, the interaction with human activity.* Following this idea, exhibition developers expanded upon and created a new narrative different from that which initially accompanied the original 32 prints, and one that more fully supported the themes found throughout the museum.

In most exhibitions, the Big Idea does not actually appear as text for the visitor, but it certainly might. *Sharks are not what you think* is a classic Big Idea from an exhibition at Monterey Bay. Because this take-home message was considered so important to the developers, it was incorporated into the exhibition's introductory space (Rand 1993). Similarly, the Big Idea for The Field Museum's *Evolving Planet* exhibition was also presented within its introductory space (Fig. 5.1). Placing the Big Idea at the entrance of an exhibition prepares the visitor for the theme of their journey, and it provides a framework for the information and objects that await them. This idea can also be repeated within the exhibition's conclusion, reminding visitors of the message and reinforcing its importance.

Exhibition topics can be approached in many different ways, but in order to not get lost in the possibilities, the Big Idea provides direction for the exhibition team, informing decisions, and—extremely important—keeping

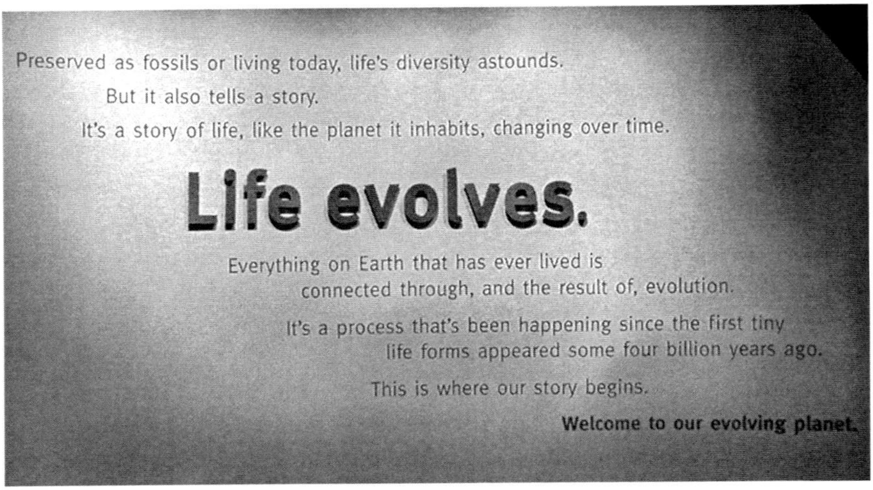

FIGURE 5.1. The Big Idea used throughout the development of The Field Museum's *Evolving Planet*—Everything on Earth that has ever lived is connected through, and the result of, evolution—is found within the exhibition's introductory text. Photo by R. Kissel.

the team focused on the scope of content delivery. It is the foundation for all of the story ideas and exhibition concepts that will be developed during the process. Any element introduced into the exhibition—be it a text panel, object, or interactive—should fall within the scope presented by the Big Idea. The Big Idea, then, is a valid test for any and all suggested elements during the planning phase of the exhibition. Always ask, "Does this object, interactive, discussion, etc. support the Big Idea?" Establishing a Big Idea at the onset of exhibition development helps ensure that 1) the exhibition maintains a clear focus; and 2) the final product will possess a singular message that is clear to visitors, creating a cohesive story and more satisfying visitor experience for many. All parts of an exhibition complete and add up to the Big Idea.

Learning and Experience Goals

After an exhibition team has established the Big Idea, it can and should develop a series of exhibition goals. While developing and applying a Big Idea ensures that a singular theme is present throughout the exhibition, defining both learning and experience goals, provides an even deeper level of guidance. Building on the foundation that is the Big Idea, the goals are the floor upon which the exhibition's content and design are constructed. Considering goals takes a team beyond *just* interpretation to thinking about the exhibition as a three-dimensional, sensory experience that engages visitors in multiple ways. It is the heart of visitor-centered design—this direction follows Weil's (1999) charge to produce exhibitions that are not *about something* but are *for somebody*.

Learning goals outline the concepts and content that you'd like visitors to take away from their visit. What do you want to hear them say they learned? What are the main messages? The Big Idea provides a central concept for the exhibition; the goals are extensions of this concept. For a traveling exhibition on maize developed by the Museum of the Earth and Cornell University, the finalized Big Idea was *We can study evolution by looking at maize domestication.* From this Big Idea, the established learning goals were:

- Visitors will understand the cultural and evolutionary history of maize.
- Visitors will understand the relationship between genetic diversity and evolution.
- Visitors will understand how scientists apply evolution (including knowledge of genetic diversity) to crop improvement.

Learning goals can help shape major content areas within the exhibition, and they focus on the major ideas that exhibition developers would like visitors to take away with them.

Experience goals reference the methods of visitor engagement and the experiences found within the exhibition. Will visitors collaborate, examine, and discuss? How and what will visitors feel as they explore the exhibition: surprise, wonder, or curiosity? A traveling exhibition developed by the New York Hall of Science—*Charlie and Kiwi's Evolutionary Adventure*—presents an introduction to biological evolution, targeting visitors aged seven and older. As noted on the exhibition's website, the exhibition was developed and designed within the framework of the following experience goals:

- Visitors will see evidence of evolutionary connections between dinosaurs and birds.
- Visitors will enjoy the story of Charlie, whose curiosity leads to understanding of how evolution works.
- Visitors will engage in activities about the key evolutionary concepts of variation, inheritance, selection, time and adaptation.

As written, these goals describe the desired actions of the exhibition's visitors. Describing and defining the desired visitor experiences will help guide the development and design of the exhibition and its elements, from label copy to interactives.

Like learning goals, experience goals directly relate to and support the exhibition's Big Idea. If clearly articulated goals are not established, determining the success of an exhibition is difficult. Goals are also a measure of whether or not an exhibition is reaching its full potential. In the Standards for Museum Exhibitions and Indicators of Excellence (Professional

Networks Council of the American Alliance of Museums 2012), evaluation with respect to an exhibition's goals is listed as one of seven major standards: "Evaluation studies are conducted during development and/or after opening the exhibition to understand its impact on audiences in relation to the project's goals" (ibid., p. 3).

Consequently, both learning and experience goals have a direct influence on a team's exhibition development strategy. Each element should be labored over for its contribution to the overall intent of the exhibition. For example, adding an interactive merely for the sake of having an interactive element might not further an exhibition's goals. All conceived elements should be measured against the Big Idea and goals. The creation of the Big Idea and the goals ideally occurs very early in the exhibition development process, as they guide the initial content research and subsequent design philosophy. Their writing is primarily a collaborative process among the developers, curators, and education specialists. Following this initial groundwork, the exhibition team has focus and direction for the remainder of the exhibition design process.

Bubble Diagrams

With an established Big Idea and a core set of learning and experience goals, creating the basic outline of an exhibition's content—or more properly, its narrative—may follow. And in that statement, the term *may* is very specifically defined as an expression of permission. Developing an outline of an exhibition's story before a Big Idea and concrete goals are developed can lead to either an unfocused final product or, upon the realization that the trajectory of the exhibition's creation is confused, much wasted time and repeated work by the team.

Despite its usefulness, as just described, it is important to realize that an outline in the traditional sense is not necessarily the best product for this stage of exhibition development. A traditional outline forces thoughts into a linear arrangement with no overlap allowed. This organization might work for some exhibitions, but it certainly will not work for all. Instead of an outline, then, it is usually better to produce a Bubble Diagram. Following the description provided by McLean (1993, 60), bubble diagrams are "idea maps" that "encourage visual thinking by linking subjects and ideas in space. The size of each 'bubble' establishes emphasis; the relations among bubbles establish a rough layout of space and provide the foundation for developing a floor plan." Grasso and Morrison (1999, 173) provide a similar description, describing a bubble diagram as a "schematic representation of the sequence and relative importance of ideas and experiences." But, again, keep in mind that ensuring a linear or sequential visitor flow might not factor into every exhibition.

The creation of a bubble diagram is an important step of the exhibition development process; it is the moment when the exhibition developers—working with the exhibition's curator(s) or content specialist(s)—develop not only the concepts, or topics, within the exhibition but also their relative importance. However, while the Big Idea and learning and experience goals are, and should remain, fixed, the bubble diagram can, and likely will, evolve as development continues into its next phases. Embrace the creative process that travels hand-in-hand with further content research. And think of different ways to organize content. The bubble diagram connects thoughts and concepts—shapes and colors—to create a thematic framework upon which writing and design will be built; it illustrates the general storyline of the exhibition, and it can introduce ideas about desired visitor circulation.

In 2013, *Raising the Dead: The Art and Science of John Gurche* opened at Ithaca's Museum of the Earth. For the exhibition, which explored the intersection of art and science through the work of renowned paleoartist John Gurche, the bubble diagram divided the space into three main sections: the introduction; a general history of paleoart, including prehistory in pop culture; and the primary story on Gurche, which contained three subsections (Fig. 5.2). In this example, the Bubble Diagram is mapped to the actual physical space of the exhibition, but this mapping—while potentially useful—is not always necessary. For the Cayuga Nature Center's *Busy Bee: Life in a Honeybee Hive* exhibition, the diagram takes no consideration of the space, focusing solely on the exhibition's themes and their relationships (Fig. 5.3)

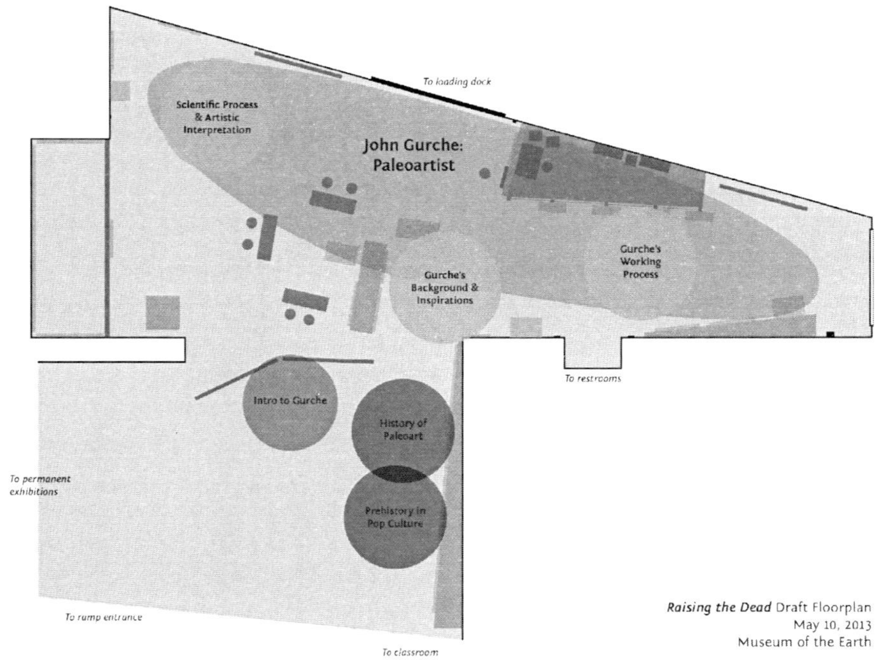

FIGURE 5.2. The bubble diagram for Museum of the Earth's exhibition of artist John Gurche's work incorporates consideration of the exhibition's physical space. Courtesy Beth Stricker, Paleontological Research Institution.

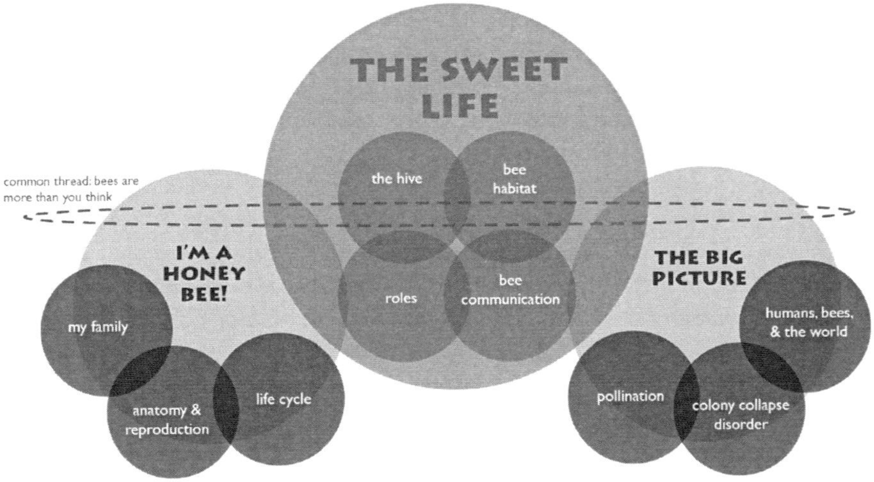

FIGURE 5.3. The simple but effective bubble diagram from Cayuga Nature Center's exhibition on bees; unlike the example in Figure 5.2, this bubble diagram is more conceptual, with no consideration of space. Courtesy Beth Stricker, Paleontological Research Institution.

As illustrated by these examples, bubble diagrams are an important and logical step in the early development process. Remember, all topics are bottomless. A concise direction must therefore be developed in order to create focus for an exhibition team. The bubble diagram presents that necessary, overall thematic structure. It is a concept map for the general organization of an exhibition's themes and content, serving as a reference to help create a narrative that informs the writing of label copy.

Conclusion

Creating exhibitions for the museum audience requires the consideration of not only content but also the visitor. As outlined by Falk (2009) and Chapter Four, the museum visitor experience is a personally constructed entity. Visitors to the same exhibition bring with them unique interests, experiences, values, and prior knowledge. Even the same visitor is likely to have different experiences within the same exhibition upon subsequent visits. Museum visitor experiences therefore vary from individual to individual and even from visit to visit. Clarity of message, then, is of the utmost importance. Crafting a concise Big Idea and establishing a set of learning and experience goals provides the clear direction required of any successful development process. Without them, the enormity of the content associated with *any* subject can quickly overwhelm and lead to an unfocused assemblage of themes and topics that—together—don't produce a cohesive narrative. A clearly told story cuts across visitor typology.

CHAPTER SIX

Strategies for Encouraging Engagement

"When you make the finding yourself—even if you're the last person on Earth to see the light—you'll never forget it."

—Carl Sagan

Now that the Big Idea and exhibition goals have been finalized and the Bubble Diagram produced, a logical next step is to develop effective methods for content delivery. As McLean suggests (1993, 102), "exhibits should be like good conversations"; they should not "lecture at" visitors but instead "engage them in dialog."

Museums began moving beyond didactic panels as early as 1889, when the Urania opened in Berlin. This early model of a science center displayed what Pallo (2009, 171) described as "devices and models of machines that could be used by the visitor." Today's museums employ a variety of techniques, which range from large-scale, 4D immersive story theaters (Mouw and Spock 2007) to low-tech panels that slide or lift to reveal text or images (McLean 1993). The key to deciding which way to engage audiences is to let the content lead the process. What works best is conditional, based on the specifics of an institution—its mission, audience, budget—and, importantly, the Big Idea.

There are many examples of fabulous high-tech engagement strategies—there is no shortage of cool gadgets and mind-bending technology out there—but these solutions pose significant challenges: they are often admittedly expensive and can require industry expertise to develop, install, and maintain. And let's face it, most institutions have very limited exhibition budgets and are looking for ways to increase the bang for their buck! It is easy to get wrapped up in the hype of new technology—it is often visually attractive and very appealing—but as Crean (2001, 214) correctly notes, "The most important directive to remember when planning...is not to let technology drive the process. The communication objectives of the exhibition should be first priority."

It is important to heed such warnings and to not let the technology run away with the message, as "it is particularly easy to bombard visitors with so many stimuli that they have little capacity left to attend to the educational message" (Perry 2012, 100). Technology should be integrated into an exhibition space and user experience, and it is better to "use a little technology well, not a lot of technology poorly" (Wyman et al. 2011, 465). So, with this in mind, let us look at a few possibilities and considerations for dealing with each.

Is it Hands-on, Interactive, or Participatory?

There are many different delivery vehicles or strategies that can be employed for exhibitions, from didactic text panels to elaborate story theaters. But before we begin to explore the variety of options in any depth, we must first acknowledge some of the limitations of our terminology. Museum professionals have gone to great pains to categorize visitor engagement, and it is easy to get consumed in the semantics of it all. Terms like *interactive*, *participatory*, and *hands-on* have all been used to describe different ways of engaging visitors (e.g., McLean 1993; Office of Policy and Analysis 2002a; Humphrey and Gutwill 2005; Simon 2009).

While definitions can provide the framework for thinking about individual elements, they can also be limiting, as they tend to privilege structure and design over content and experience. Instead, a much more organic process is recommended, wherein the strategies of engagement are developed within the context of the overall interpretative framework and grow from the content, not the other way around. This approach is why we start with the Big Idea and exhibition goals—if you don't know what you are trying to do, how are you supposed to do it? It is only then that an exhibition team may move on to considering strategies of engagement. Think of hands-on, interactive, and participatory as variations on a theme, and at its heart is communication. So, ultimately, in the words of McLean (1993, 102): How is the exhibition going to create a "dialog" with the visitor?

Hands-on

Are hands-on displays interactive? McLean (1993) makes the point that interactivity is about being reciprocal, which she suggests would disqualify many hands-on displays from this conversation. Kennedy (1990, 2) describes hands-on displays as "exhibits that require some kind of physical interaction between the visitor and the exhibit," which in his case includes everything from simple lift-and-drops to complex technology based multidimensional activities. Kennedy (1990), like McLean, also makes a distinction between "hands-on" and interactive or participatory elements (discussed later), which he suggests do not necessitate physical contact. Haywood and Cairns (2006, 2), on the other hand, contend that "interactive exhibits allow for interaction in some form beyond mere visual perception." As we move forward, we suggest that this debate—"Is hands-on interactive?"—is a moot point. When content and message lead the

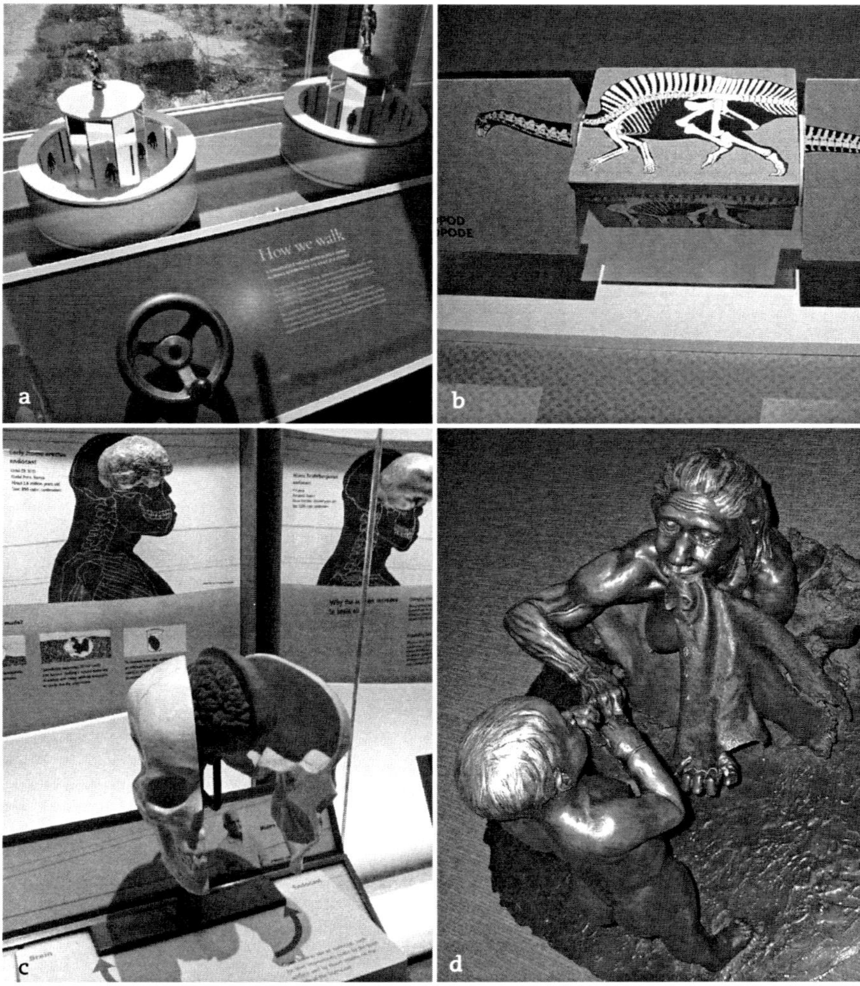

FIGURE 6.1. Simple but very effective mechanical interactives: (a) zoetrope at the Natural History Museum of Los Angeles County, showing differences between the walks of chimpanzees (left) and humans; (b) spinning, three-part drum illustrating the diversity of dinosaur body types at the Royal Ontario Museum; (c) hinged skull cast at the National Museum of Natural History, demonstrating the difference between the brain and an endocast; and (d) touchable, bronze Neanderthal woman and child reconstructions by artist John Gurche. Photos by R. Kissel (a), (b), and (c) and S. Chicone (d).

design, an exhibition team will focus on the best strategies to maximize the message; while in some cases this might be a simple lift-and-drop panel or other mechanical interactive (Fig. 6.1), in others it might take the form of a complex intellectual interaction. What category it ultimately fits into is not as nearly as important as its effectiveness as a strategy for relaying the content and, therefore, supporting and serving the exhibition's goals and audience.

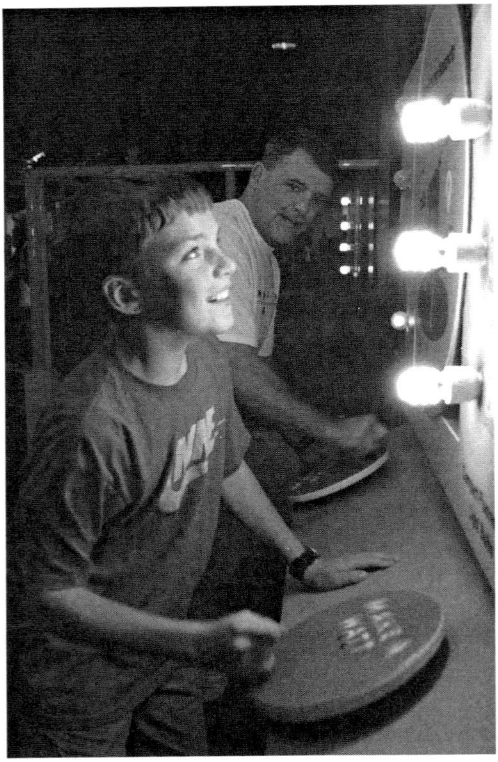

FIGURE 6.2. A young visitor spins a wheel to generate electricity at the *So Watt!* exhibition at the Reuben H. Fleet Science Center. Photo courtesy of The Reuben H. Fleet Science Center.

All this being said, we do recognize that different delivery methods result in different types and levels of engagement. Consider the distinctions between what Humphrey and Gutwill (2005) describe as *Planned Discovery* (PD) and *Active Prolonged Engagement* (APE) displays. In a PD display, "the museum has 'planned' the exhibits so that visitors will 'discover' the concepts" (ibid., 7). For example, a visitor approaches a display, manipulates it, and is rewarded with a specific outcome. Turn the crank, watch the turbine spin to create energy, and then see the light bulb attached to the turbine glow as a result (Fig. 6.2). Active Prolonged Engagement (APE) is a way of thinking that suggests, "***Active*** visitors are in the driver's seat, deciding for themselves what to try next rather than following a set of instructions from the museum. ***Prolonged***...visitors spend more time with the exhibits, getting more involved with the phenomena than they might at other exhibits. And ***Engaged*** visitors try a variety of things at the exhibit, with each action they take somehow building on their previous actions" (APE Project, n.d., para 1). APE envisions an important shift in authority from the museum to the visitor—a shift "from planned to self discovery" (Humphrey and Gutwill 2005, 2), characterized by exploration, investigation, observation, and construction. These types of displays, pioneered at the Exploratorium in San Francisco, are open-ended and result in multiple outcomes and engagements. They also replicate a scientific investigation, reinforcing the process of science. But this isn't to say that they are the best strategy for *every* context; sometimes all you need is a glowing light bulb.

Participatory

The two approaches described in the preceding section, PD and APE, are still different from what Nina Simon (2009, para 5) describes as participatory engagement; she defines a "participatory cultural institution as a place where visitors can create, share, and connect with each other around content." She insists that, "instead of being 'about' something or 'for' someone, participatory institutions are created and managed 'with' visitors" (ibid., para 6). This approach has taken the form of talk-back walls (Carnegie Museum of Natural History and the California Academy of Science), video kiosks (*Voices for the Lake* at the ECHO Lake Aquarium and Science Center), photograph and gallery tagging exercises (Natural History Museum of Los Angeles County), and blogging from within the exhibitions (*Month at the Museum* at Chicago's Museum of Science and Industry).

In examining participatory engagement strategies, Simon (2009) outlines two essential components needed to design for participation: 1) understand your audience, and 2) have clear institutional goals in mind. Sound familiar? The approach to audience in this particular case is slightly different from what was presented in Chapter Four. Instead, Simon's use of Forrester Research's social technographics profile groups social media audiences into the following six categories (Simon 2009, 11):

- *Creators*, people who produce content, upload videos, write blogs;
- *Critics*, people who submit reviews, rate content, and comment on social media sites;
- *Collectors*, people who collect links and aggregate content for personal or social consumption;
- *Joiners*, people who join social networking sites like Facebook and LinkedIn;
- *Spectators*, people who read blogs, watch YouTube videos, and visit social sites;
- *Inactives*, people who don't visit social sites.

Simon suggests that creators actually comprise only a small part of the landscape. Given the presence of six categories, why do exhibitions seem to consistently single out creators? As it stands, museums tend to target creators with their participatory strategies, yet, as Simon (2009, 13) suggests, "exhibitions that invite self-expression appeal to a tiny percentage of museum audiences. Fewer than one percent of the users of most social Web platforms create original content. Would you design an interactive exhibit that only 1% of visitors would want to use?"

Creating opportunities for "open-ended self-expression" (e.g., StoryCorps) only targets a small number of visitors. If a more constrained experience and a menu of opportunities are offered—think collaborative, not just open ended—participation will likely increase. The key is to "use intentional design to set a stage for participation that is clear and comfortable" (Simon 2009, 14). And this isn't just about creating participatory experiences for the visitor; as with all potential strategies, it is about choosing the right tool for the job—not just designing something to be "participatory" for its own sake, or using "technology" for its own sake, but letting the content drive the medium chosen to deliver it. "The experience isn't solely about providing value to the participants. It's also about providing value to their critics and spectators, which include museum staff" (ibid., 16), and to the institution.

Multimedia

The different ways to engage visitors described in the preceding section can be realized through a variety of delivery methods, including multimedia displays. So, then, what does it mean to be multimedia? Maria Piacente (2001, 222) defines it as "any audio-visual, computer, photographic, video or live theater program or production." This admittedly broad definition could incorporate nearly every strategy that includes technology in some form; indeed, Dirbin and Rickhoff (2010, 221), in reference to the National Research Council's report on Informal Science Education—*Learning Science in Informal Environments: Places, People, and Pursuits*—acknowledge that the term media "covers a broad spectrum of content delivery methods." Be mindful that "museums will be most effective when they use electronic technology in ways that accentuate their unique role as three-dimensional, geographically located places that bring together real people and real things" (Mouw and Spock 2007, 46). In the interest of clarity, this approach can be broken down to include more manageable divisions and highlight several emerging trends, including gaming, mobile, and immersive environments.

- *Gaming*. Play is increasingly recognized as a legitimate learning activity (Spock 2004; Adams et al. 2004; Mitleer et. al. 2007; Bushara 2009), and gaming—though rule-based and not open-ended, which is often used to describe other forms of play—is recognized as falling within this arena (Danks et al. 2007; Edwards and Schaller 2007; Beale 2011). Gaming strategies appeal to digital natives and take advantage of existing technologies, including visitors' smart phones within the museum walls and the online gaming community outside of it (Birchall et al. 2012; Fabrikant 2012). As with each strategy discussed, and at the risk of over repeating the sentiment, remember that content must lead the design. And it is no different with gaming: "Museums must carefully conceptualize their goals in order to choose the appropriate game model" (Edwards and Scahller 2007, 105). Most museum games fall within the "casual but purposeful" category (Birchall et al. 2012). Remain true to the content—this is a given—but also remain true to the strategy. For example, it isn't enough to take an existing experience and make it seem like game. Be careful not to gamify the experiences; in other words, do not take the "dynamics and player motivations inherent in game playing and us[e] them to make something that isn't a game seem like one" (Rodley 2011). The strategy works best when both the content and the medium are honored.

- *Mobile.* Mobile devices are increasingly the most common way for individuals to access the Internet. This, combined with their ubiquitous nature, makes them a key means of visitor engagement. They provide "up to the minute contextual information" (Johnson et al. 2010, 17) and are particularly useful for "tying content to location or taking an experience outside of the confines of an institution" (ibid., 10). The 2010 Horizon Report on Museums (Johnson et al. 2010, 10) suggests that mobiles "can be used to encourage deeper connections among individuals, institutions, and collections. The range of technologies converging in mobile devices is very broad, as is the variety of ways they can be applied: GPS and compasses allow sophisticated location and positioning; accelerometers and motion sensors enable completely new forms of control and interaction; digital capture and editing bring rich tools for high-resolution video, audio, and image capture and even editing—more and more, mobiles encompass it all."

While the development of some mobile applications requires a sophisticated technological skill set and has the potential to incur significant costs, the many open-sourced web-based platforms have made it easier to develop in-house audio/cell phone tours, pod casts, apps, code readers, and even such seemingly high-tech applications as augmented reality and unique location-based services. A number of web-based resources can help a team navigate the steps of planning and development. These types of technologies are surprisingly accessible, even to the small museum, through partnerships and a little legwork. Of course, the use of personal electronic devices automatically discriminates against those visitors not possessing such technology. And if the presentation and appreciation of the object—specimens and artifacts—is key to the museum and its mission, then the incorporation of visitors' personal technology might lead to a museum experience that is similar to the 'everyday' and not unique to the museum context. The incorporation of such technology certainly warrants considerable discussion within any institution.

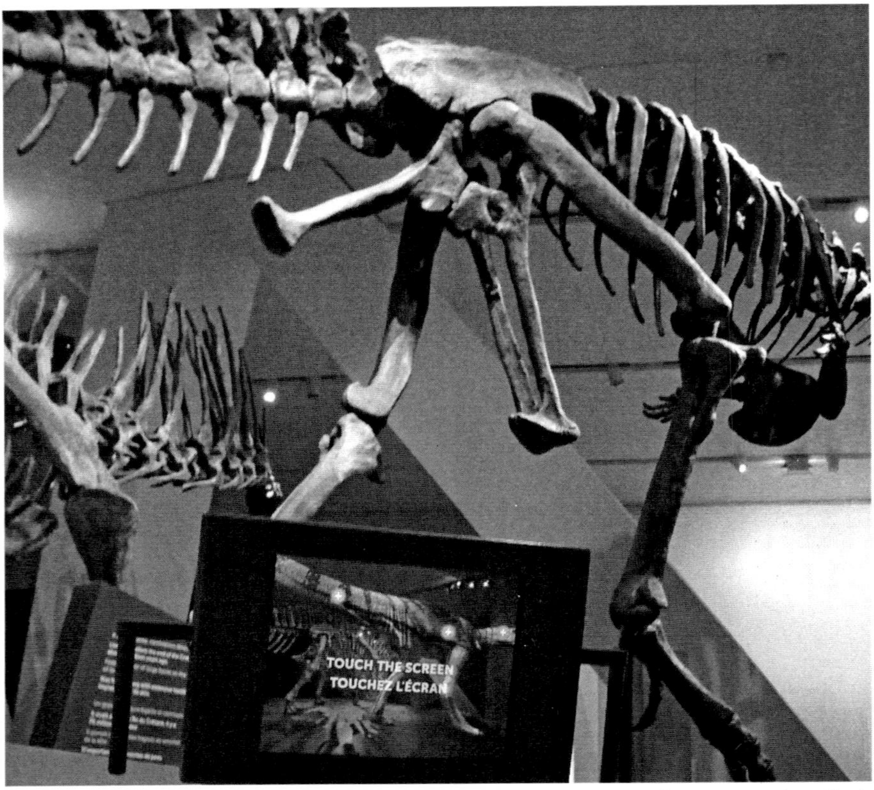

FIGURE 6.3. The Royal Ontario Museum's *Ultimate Dinosaurs* used augmented reality to bring dusty bones to life, with: a) three permanent AR stations, and b) a downloaded app for mobile devices that activated additional experiences at designated markers. The app also played a role in the exhibition's marketing campaign, with outdoor posters and other materials featuring the markers. Photo by R. Kissel.

Augmented reality adds additional layers of content and interpretation by using either visual marker cues, or positional data (Johnson et al. 2010, 16). This technology can provide superimposed environments, annotated landscapes, or spatial acoustic overlays—it literally "augments" or "adds to" reality. It enhances what visitors can perceive with their senses, helping them see the world around them in new ways (Johnson et al. 2012). For example, in the Royal Ontario Museum's *Ultimate Dinosaurs: Giants from Gondwana* exhibition, visitors can witness a full-sized, fully fleshed dinosaur, right where the skeleton is standing (Fig. 6.3), or at the Smithsonian's Museum of Natural History visitors can make themselves into a Neanderthal (MEandethal) (http://itacitus.org/). Location-based services (Johnson et al. 2010, 20) use user location to deliver tailored content. They can take the form of geocaching, for example, where a "museum can 'pin' information to a given object or gallery location and have it pushed to a user once he or she has reached that location" (ibid., 21).

- *Immersive Environments.* Immersive environments run the spectrum from the physical to the virtual and everything in-between, combining the most effective of each strategy to maximize visitor engagement with exhibition content. As Lorentz (2006, 42) suggests, "people have different perceptions of reality based on various developmental, environmental and cultural experiences, and this needs to be taken into consideration"; it is "the totality of the planned experience that makes it immersive."

Dioramas

No discussion of science-based exhibitions and engagement strategies would be complete without mention of the diorama—that celebrated fixture that dominates many childhood memories. Dioramas transport visitors to a different place and time—when our early hominin ancestors struggled for survival, or when bison roamed the vast open prairies (Fig. 6.4). Dioramas began appearing in museums in the late 19th century, as institutions moved away from static displays of solitary taxidermied specimens—still evidenced in the nineteenth-century arrangement preserved in the *Great Mammal Hall* of the Harvard Museum of Natural History in Cambridge, MA (Fig. 6.5). It is often Olof Gylling, a trained scientist and taxidermist, who is credited with co-opting the term *diorama* to describe the habitat scenes appearing in Sweden and the United States in the late 19th century (Hoyle, 2008). The term diorama literally means "through that which is seen" or "to see through" and was first used by Louis-Jacques-Mandé Daguerre, of Daguerreotype fame, to describe his patented theatrical technique.

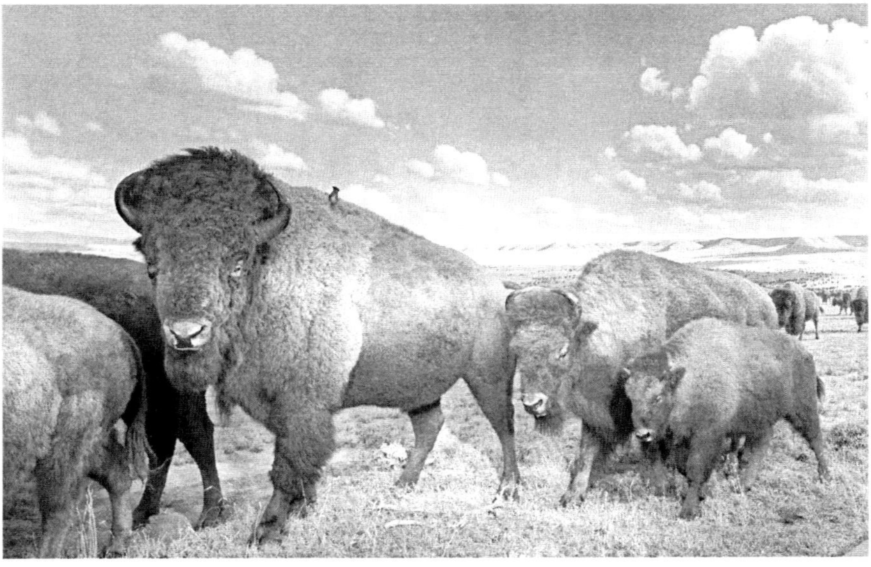

FIGURE 6.4. Diorama of American bison at the American Museum of Natural History. Image # 318958 American Museum of Natural History Library.

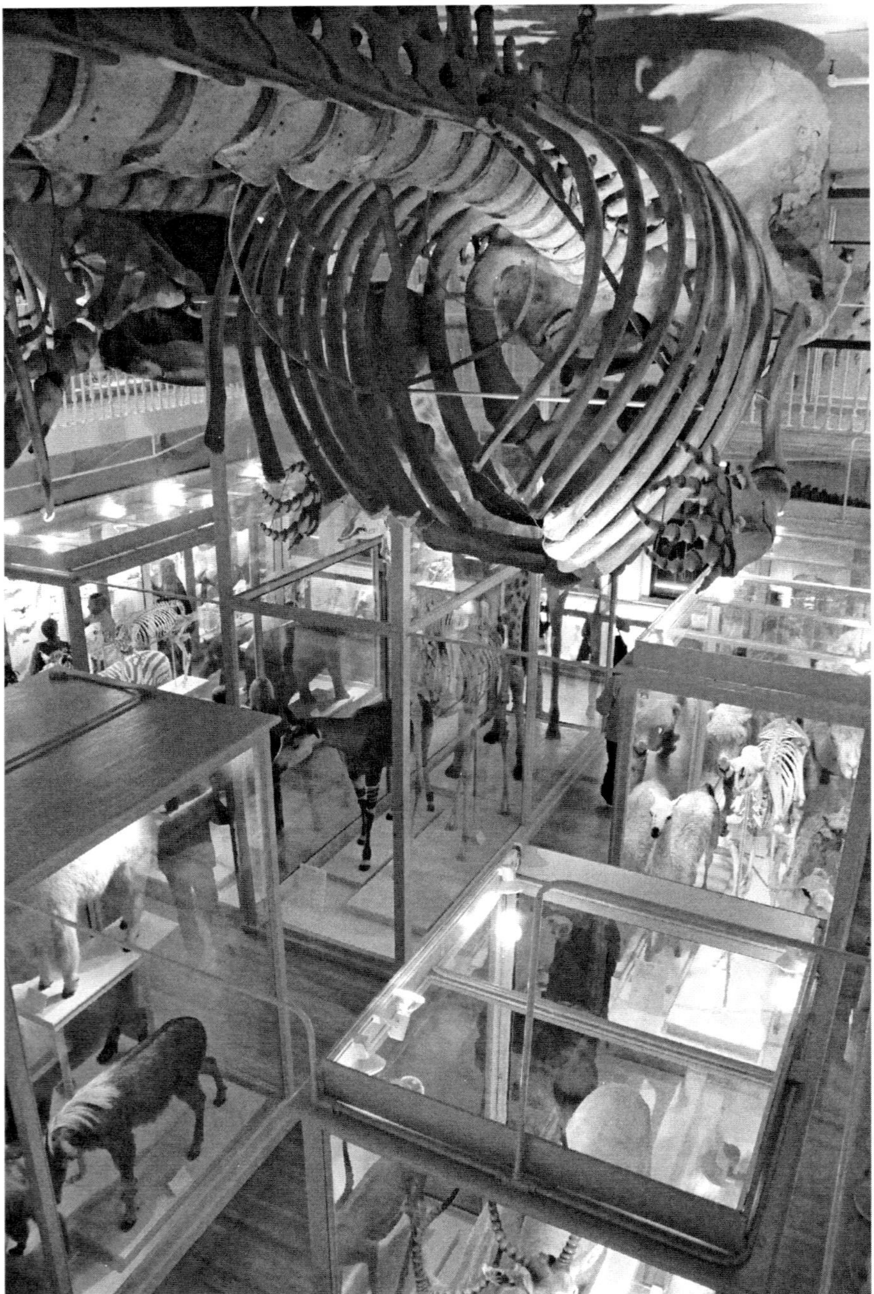

FIGURE 6.5. A view of the *Great Mammal Hall* of the Harvard Museum of Natural History. The original gallery dates back to 1872; in honor of the 150th anniversary of the founding of Harvard's Museum of Comparative Zoology (MCZ), it was renovated to its original look and feel, reflecting the vision of MCZ founder Louis Agassiz. Courtesy Harvard Museum of Natural History, photo by Audrey Anderson, Harvard Crimson.

Dioramas, both ethnographic and biological (Haraway 1990), helped conserve (read preserve) culture and habitat in "what was thought to be their most pristine form" (Harrison 2005, 36). But while these classic vignettes have some obvious benefits for displaying global biodiversity, as an approach they are not without valid critique (Harrison 2005). Ethnographic dioramas can present "static, dehumanized caricatures that fuel classic and colonial stereotypes," and representations of "the supposedly 'primitive' technologies of non-industrialized peoples" (Steiner 1991, 34). These are real concerns with problematized histories. The University of Michigan's Museum of Natural History worked with the Native American community to ultimately remove 14 dioramas from permanent display. So, is the diorama as an exhibition strategy "best abandoned," or as Harrison contends (2005, 35), can it be linked to "life and not death"?

In spite of the valid critique levied on their representational impact, we need to be careful not to throw the baby out with the bathwater. The diorama is still a familiar and often engaging strategy; it attracts visitors' attention, is visually appealing, and often central to a gallery layout (Harrison 2005). Visitor studies have attempted to pin down the efficacy of this approach (Bitgood 1996; Reiss and Tunnicliffe 2001; Lorentz 2006) with varying degrees of success. What we do know is that visitors enjoy and appreciate them. As a result, institutions have invested millions of dollars to renovate them (e.g., Boston Museum of Science, American Museum of Natural History, Carnegie Museum of Natural History, Denver Museum of Nature and Science, California Academy of Sciences), and they offer more than an aesthetic experience. In fact, it has been argued that they are a "valuable tool for the learning of biology and development of biological interest" (Reiss and Tunnicliffee 2011, 448). Dioramas can be used to develop observational and interpretive skills through the construction of narratives, and "inquiry based science...can be encouraged and developed at dioramas by emphasizing observation and constructing meaning from them" (ibid., 450). They have pedagogical functionality.

While a glass wall separated the visitor from the vignettes displayed in the first fully developed large-scale dioramas of the 20th century, this fourth wall was later re-envisioned and dioramas took the form of open tableaus effectively drawing the visitor into the scene (Reiss and Tunnicliffe 2011). Visitors can walk into a glacier (Fig. 6.6, Rochester Museum & Science Center), among the trees of a 310-million-year-old forest (Burpee Museum of Natural History), or through the rooms of a recreated Egyptian tomb (The Field Museum). Today, immersive environments expand that experience to include 3D and even 4D elements, with theatrical and ambient sound, animation, and even kinetic experiences of smell, temperature, and touch, breathing new life into the once static, frozen tableaus. Reactive walls that make use of body-scanning technology, like those used in the Royal Ontario Museum's *Ultimate Dinosaurs* exhibition, showcase projections (in this case, dinosaurs) that react to visitors' movements.

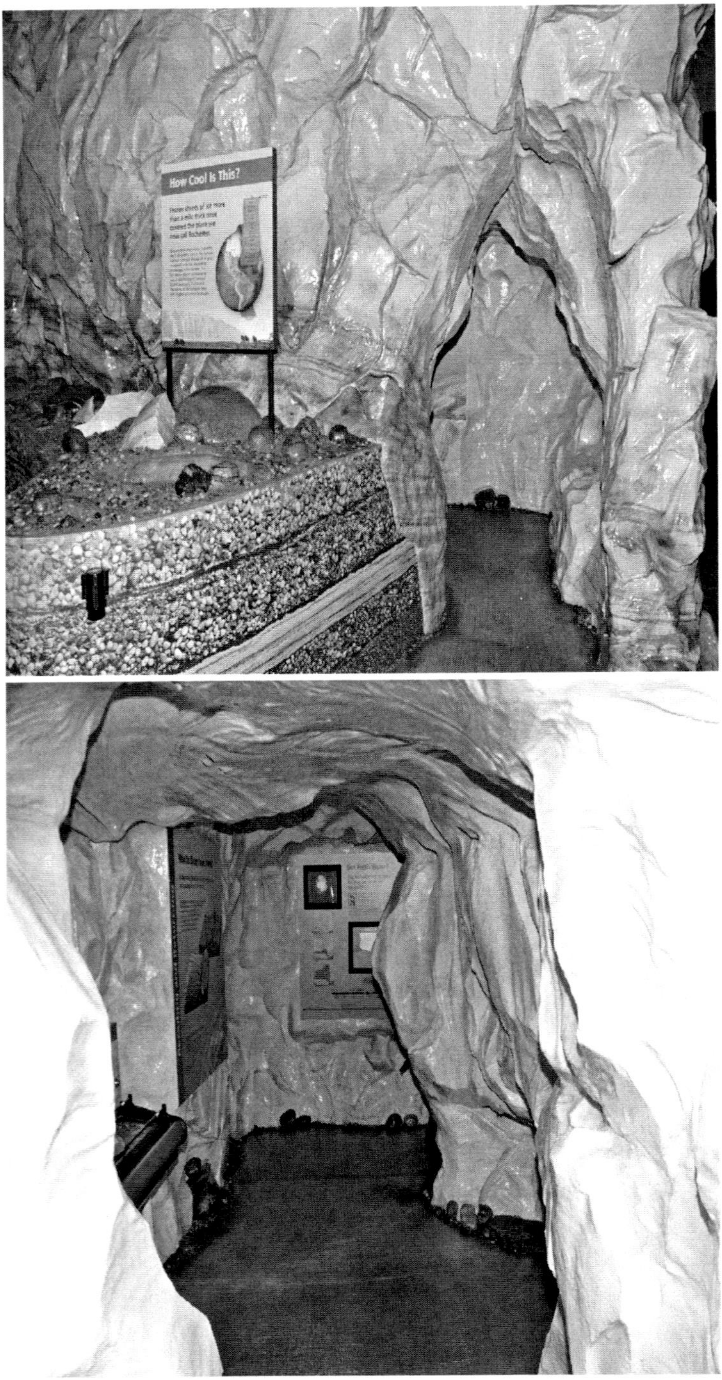

FIGURE 6.6. Constructed environments like the walk-through glacier at the Rochester Museum & Science Center offer a more immersive experience for visitors. Photos by Vincent Chicone.

From the early 20th century dioramas at the American Museum of Natural History to immersive objects like the *Maori Meeting House* at The Field Museum, the recreated vignettes like *A Pequot Village* at the Mashantucket Pequot Museum & Research Center, the 4D immersion theaters like at the National Aquarium in Baltimore, and the object/story theaters like those produced for a range of museums by Science North, the unifying element is the central role of narrative, whether implicit or explicit. This point is important because "stories are part of how we learn" (Reiss and Tunnicliffe 2011, 456). Explicit narrative embeds the visitor into a storied experience, whereas the "presentation of implicit narrative encourages visitors to recall or articulate their own narratives, leading to a dialogue—internal or social" (ibid., 455).

Design and Logistics

What we have covered so far are brief descriptions of some examples of the range of engagement strategies used in exhibitions. Despite our enthusiasm for the untapped potential of many of the emerging and traditional techniques, we encourage you not to get starry eyed and be blinded by the technology. Be mindful that technology for technology's sake is just that: technology, not engagement. While we have touched briefly on the *who* and the *what*, we have neglected, up until now, to address the *how*. We should not, and for that matter cannot, consider these different engagement strategies without also considering their design. Beyond clear goals and intended outcomes led by the exhibition team's Big Idea and fitting within the larger interpretative framework, there are other logistical considerations and concerns. Many of the strategies presented in the preceding sections are individually impressive (Who wouldn't want to experience the games at the Colosseum in Rome while standing squarely in its 21st century ruins, for example?), but what does it actually take to implement these ideas? And beyond that, what does it take to maintain them?

Just as we started this whole process with a unifying Big Idea and a set of goals, both learning and experiential (Chapter Five), we need to approach each strategy with a concept and goals. To echo Adams et al. (2004, 166), "Think about the content and process before you think about the delivery system." What do we want people to learn in this particular display element and what do we want people to experience? What do you want them to do and what do you want them to feel? From that viewpoint, the exhibition team can consider the options. Would a straightforward text panel work best, for example, or does the content call for something more than didactic text? Considerations include:

- *Your points of engagement should be "self-evident."* There are only a few seconds to capture a visitor's attention (Adams et. al. 2004; Samis 2007; Humphrey and Gutwill 2005), so, whether it is a lift-and-drop or a multimedia kiosk, think about how it is going to not only attract, but also sustain, the visitor's interest and attention.

- *Make sure the exhibition elements are accessible, both physically and intellectually.* Consider the seven basic principles of universal design. Visitor height variation can present significant logistical challenges, but Kennedy (1990) offers some great tips in this area. For example, a table 30 inches from floor to underside accommodates a standard wheelchair; a 10-inch forward-reach depth puts controls in reach of both children and seated adults; and a 20-degree angled panel makes instructions easier to read, accommodating both standing and seated visitors. See Chapter Nine for a more comprehensive discussion of Accessibility and Universal Design.
- *Consider multisided and multiuser experiences.* Museum visits are increasingly being recognized for their social benefits, within and between peer groups (Falk 2009). By making experiences that can accommodate multiple users, an exhibition element can encourage group interaction (Borun and Dristas 1997). Remember, "crowd size is an important factor in the quality and quantity of visitor engagement" (Adams et al. 2004, 161). For a large institution that can easily see hundreds of visitors through its galleries every day, a single user APE experience might not be the best use of limited exhibition real-estate; furthermore, a popular multiuser experience should be positioned away from the main visitor flow, in order to prevent a bottle neck within the space.
- *Consider multi-modal and multi-outcome experiences* as engagement techniques that appeal to different learning styles and expertise (Borun and Dristas 1997).
- *Make sure engagement strategies are relevant,* not just to the content but also to the visitor. Link the content through the experience to the visitor's "existing knowledge and experience" (Borun and Dristas 1997).
- *Prototype—prototype—prototype.* The best way to ensure that an exhibition team is headed in the right direction is to prototype its different display elements; this evaluation is not limited to mechanical components but should also include instructions and gallery placement. At its most basic, prototyping is testing the idea to see if it works. In order to simplify the process, follow these five simple steps outlined by Paul Orseli (2011): 1) *"figure out what you want to find out;"* 2) *"get out your junk;"* You will be amazed at what you can build out of some cardboard, permanent markers, and duct tape. The more rudimentary your prototype looks, the more likely visitors will recognize it as an opportunity to help in the design process rather than see it as an ill-conceived exhibition element with all the quality of an elementary school science fair; 3) *"start playing around with the pieces;"* 4) *"assemble, then iterate, iterate, iterate—Resist the urge to explain or over-explain your prototype. Just watch what people do (or*

don't do!) with the exhibit component(s). Take lots of notes/pictures/video. Then take a break to change your prototype based on what you've observed and heard, and try it out again. That's called iteration;" and 5) "*Figure out what's next even if it's the trash can.*" Because of its importance, prototyping is ideally included in the project's budget.

Conclusion

This chapter offers much to think about in terms of engagement methods and strategies, from classic dioramas to the use of modern technology. Regardless of the delivery mechanism, the ultimate focus is centered on the visitor and the content. When the exhibition is designed under this consideration, success in the form of a satisfying visitor experience is near. Within science-based exhibitions, interactive elements possess the power to make seemingly more complex concepts tangible, and, more importantly, they invite personal exploration. Interactive experiences—even and especially dioramas—engage visitors with the content, leading to questions and investigation; they involve the visitor with the process of science.

Telling a Story: Words to Paper

"At best, the genre we call nature writing requires a rare mixture of scientist, philosopher, and poet."

—Edwin Way Teale

Ask any person to describe a famous novel, and you will likely be greeted with a summary of the story told on the page. Ask any person to describe a stone tool, mineral crystal, or pottery sherd, however, and you will no doubt be presented with a bulleted list of physical features: color, shape, size, texture, and luster. While these attributes are important to the story, they are not the story—just as the physical dimensions of a book, or its paper's weight or chosen font, are not the story. Story matters. And every object brings with it a compelling story. How to best tell these tales is the responsibility of the entire exhibition team. From wall color and lighting to object mounting and graphic design, an untold number of decisions by the team members directly influence an audience's interaction and engagement with an exhibition space. We discussed a wide range of different engagement strategies in Chapter Six, but for all the immersive environments, interactive kiosks, and dioramas, labels and the written word remain the core, fundamental element of every science-based exhibition. They tell the stories of the objects found within.

Audience, Label Types, and Hierarchy

No two visitors are the same. The population that visits museums is a diverse one, with different individuals of that population possessing a suite of varied needs and motivations. Even the same person carries with him or her various interests on different visits, or within different exhibition galleries during a single day (see Chapter Four). Every visitor is unique; this makes it inappropriate to design an exhibition or write a label with the construct of a "typical visitor" in mind. How, then, can an exhibition developer be expected to write labels that both reach and satisfy the majority of visitors?

Sarah J. Chicone and Richard A. Kissel, "Telling a Story: Words to Paper" in *Dinosaurs and Dioramas*, pp 81-102.

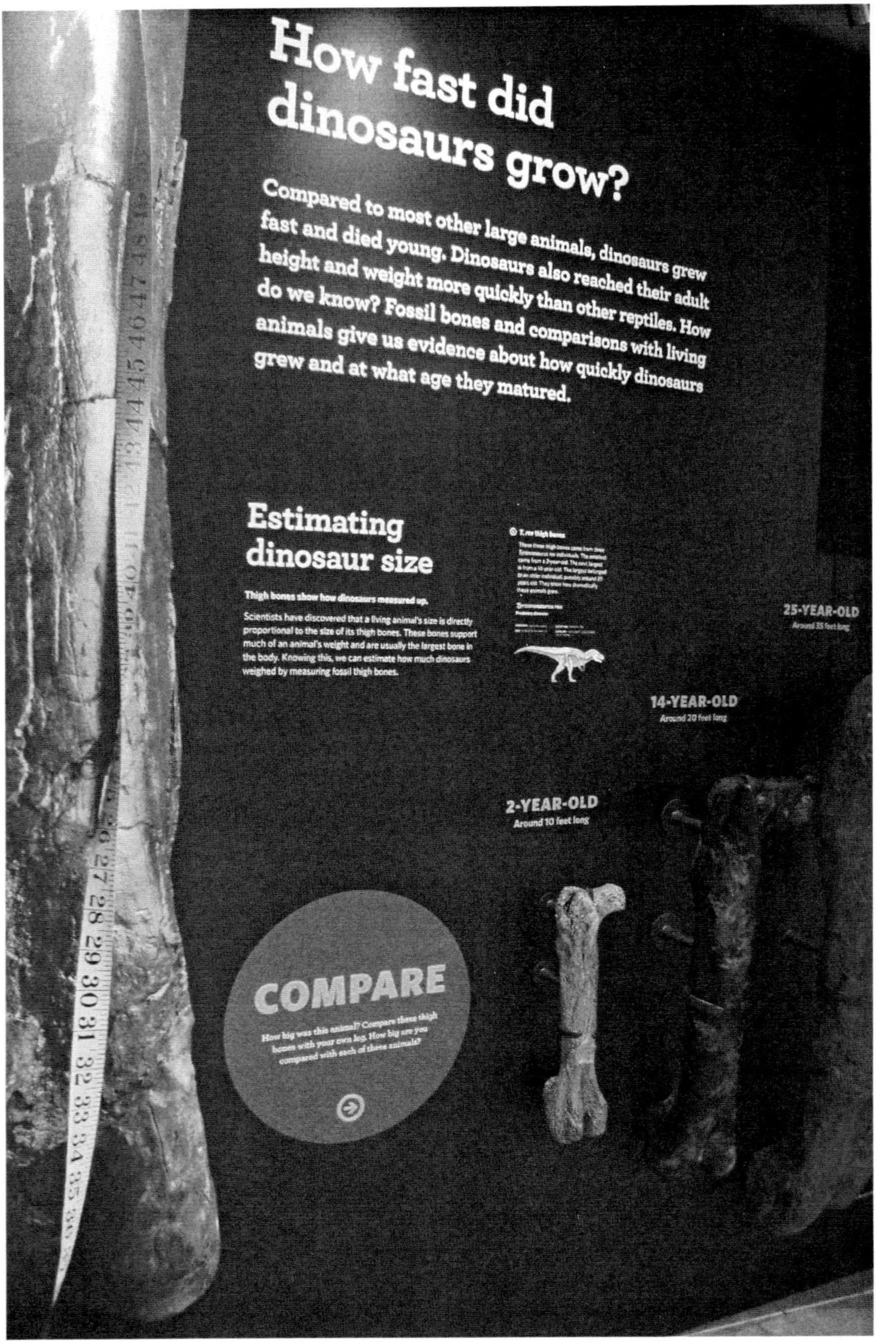

FIGURE 7.1. A display on dinosaur growth at the Natural History Museum of Los Angeles County, featuring an easy-to-understand hierarchy of labels and information. Photo by R. Kissel.

Before pencil meets paper, a number of considerations must be taken into account. For every exhibition, one initial requirement is to establish the types of labels—and their classification—that will be developed for the exhibition. As noted by Serrell (1996), there exists no universal terminology in museums to identify label types. Many institutions possess their own internal nomenclature, but regardless of the language used, the labels of any exhibition should be considered as a system defined by order and consistency. And hierarchy is key. Label hierarchy establishes linear levels of content layering and penetration within displays, and it should be obvious to the visitor (Fig. 7.1).

Serrell (1996) identifies two primary classes of labels: *noninterpretive* and *interpretive*. Examples of noninterpretive labels include object identification labels, credit panels, donor plaques, wayfinding, and prohibitive signage (e.g., please do not touch). While most noninterpretive labels are not part of an exhibition's storyline, identification labels are a critical piece of the larger narrative. They are tied to that storyline by their association with the objects that support it. Further, many visitors are first drawn to objects within an exhibition and immediately seek to satisfy the most basic and perhaps important curiosity: "What is it?" As a result of this critical role, consistency of design and presentation of information within identification labels is key. Though the information contained within identification labels varies depending on scientific discipline—biology versus archaeology, for example—standard details include: common name; scientific name[1]; age, especially for paleontological and archaeological specimens; geographic data; and the object's institution-given specimen or artifact number (Fig. 7.2).

Interpretive labels are those that carry the exhibition's narrative. They tell the story, and are developed to appeal to a wide range of visitors, communicating the depth of content associated with each label. Serrell (1996) identifies a hierarchical system composed of four types of interpretive labels—title labels, introductory or orientation labels, section or group labels, and captions.

- *Title Labels.* The highest-ranking label is that identifying the name of the exhibition. Determining the title of an exhibition—whether the exhibition covers 300 or 30,000 square feet—deserves considerable thought, as visitors' first impressions of any exhibition are driven by the title. A title that fails to capture interest or inspire curiosity is an immediate challenge to the content found inside, regardless of how engaging the rest of the labels, objects, and design are. The title also establishes tone and expectations. At Chicago's Field Museum, *Underground Adventure* is a popular exhibition that has its visitors "shrink" to the size of a penny to explore the world of soil and its creatures. As the name implies, it is a family friendly exhibition based on a surprise-filled journey, with visitors encountering earthworms, insects, and spiders that are 100 times their actual size. *Dinosaurs and Their World* at Pittsburgh's Carnegie

Museum prepares visitors for not only a parade of dinosaur skeletons but also insight into the plants, other animals, geography, and climates of their world, while the scholarly titled *Halls of Saurischian and Ornithischian Dinosaurs* at the American Museum of Natural History prepares visitors for a more academic treatment of those ancient beasts. Short, catchy, and to the point, **an ideal exhibition title should run between one and seven words.**

4 **Mastodon skeleton**
Mammut americanum
Hyde Park, Dutchess Co.,
New York
Quaternary Period
lived about 13,550 years ago
PRI 49820

Carriage
Northern China
Wood, bronze, iron
c. 1800 – 1850, Qing dynasty

This type of two-wheeled carriage, drawn by mule, horse or donkey, was the standard means of passenger transportation over land in China from the 17th to the early 20th century. Although different in style, this carriage is remarkably similar to Bronze Age chariots.

Each wheel is reinforced by 423 iron nails. This large quantity is required because the wheels have tires made from nine separate iron strakes instead of a continuous iron hoop.

2000.87.1.1–11 Gift of Joey and Toby Tanenbaum

Voiture
Chine du Nord
Bois, bronze, fer
v. 1800-1850, dynastie Qing

Ce type de voiture à deux roues, tirée par un mulet, un cheval ou un âne était le mode de transport terrestre usuel en Chine, du XVIIe jusqu'au début du XXe siècle. Bien qu'elle soit de style différent, cette voiture est remarquablement semblable aux chars de l'âge de bronze.

Chacune des roues est renforcée de 423 clous de fer. Cette grande quantité de clous est nécessaire parce que les bandages des roues sont faits de neuf sabots de fer au lieu d'un cerceau de fer continu.

2000.87.1.1–11 Don de Joey et Toby Tanenbaum

FIGURE 7.2. Object identification labels featured in the Museum of the Earth (top) and, along with a caption for the object, the Royal Ontario Museum. Photos by R. Kissel.

- *Introductory or Orientation Labels.* Like the title label, the introductory or orientation text is important for establishing the tone in the exhibition. This bit of context is arguably the most important body of text of the exhibition. Successful introductory labels are brief, as they introduce topic and scope and, as suggested in Chapter Five, may or may not include the Big Idea. *Evolving Planet* at The Field Museum is a 27,000-square-foot exhibition on the four-billion-year history of life on Earth. Its introductory text (Fig. 5.1) prepares the visitor for the exhibition, presenting the basic concept of a changing world and its life over time; the text also emphasizes the concept of evolution and establishes the exhibition as a journey that begins during the dawn of life four billion years ago. This introduction successfully draws the visitor into the exhibition, leaving the museum behind and freeing the visitor to enter and explore a new world. All in 74 words. The 15,000-square-foot *David H. Koch Hall of Human Origins* at the National Museum of Natural History successfully introduces its visitors to the six-million-year story of human evolution with a mere single sentence containing 27 words (Fig. 7.3). Every word matters, and every word is painstakingly selected. **An ideal orientation or introductory label should run between 20 and 300 words.**

- *Section or Group Labels* will account for the largest number of labels in an exhibition; they introduce topics within the individual displays. They contain a header, as well as a subheading if necessary, in addition to a body of text. The level of information falls between that presented in a title and the finer details of captions, which focus on individual objects or images. This label category, as presented by Serrell (1996), is very broad; a further division is provided in the next set of bullets. **Well crafted section or group labels range between 20 to 150 words**

- *Captions* are those labels designed to accompany very specific elements, from objects like artifacts and specimens to photographs and other images. While all objects will possess a noninterpretive identification label, not all will possess a caption; only those with an interesting story or those that strongly reinforce the narrative are generally given captions. But much like identification labels, which might be viewed before the higher-ranking labels, the information presented within a caption should function well independently while also conforming to the aesthetic and story of the other labels. **Like group labels, a word count ranging from 20 to 150 is recommended for captions.**

While the preceding label classification is common, it can be overly simplistic and we suggest a slightly more rigorous categorization, recommending that Serrell's (1996) *Group Labels* be further divided into a tiered hierarchy consisting of—from highest ranking to lowest—area introductions, theme labels, and focus labels.

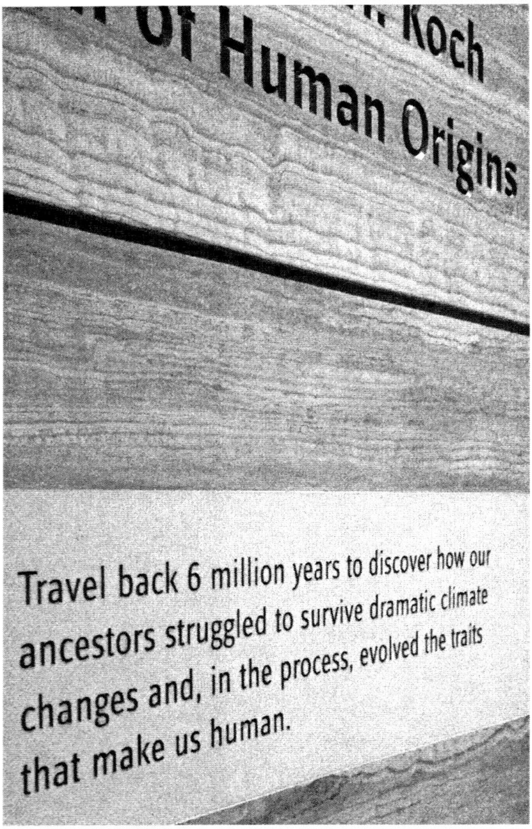

FIGURE 7.3. The entrance to the *David H. Koch Hall of Human Origins* effectively prepares visitors for a six-million-year story, all in a single sentence. Photo by R. Kissel.

- *Area Introductions.* Across their footprint, many exhibitions are divided into strict, content-related areas. An exhibition on the life of Charles Darwin, for example, might be divided into three distinct areas: Darwin's life before his voyage on the *HMS Beagle*; his adventures and studies during the voyage; and his life after the voyage. For each of these areas, a large panel that both summarizes and introduces the content ahead prepares visitors for the next section of their journey. Such previews are written with all visitors in mind, from streakers to studiers. By reading nothing but these three panels, a visitor will still have a basic understanding of Darwin's life, as well as the exhibition's Big Idea. Section previews occur at natural breaks in the storyline, and they prepare the visitor for the next "chapter" of the exhibition. At Ithaca's Museum of the Earth, for example, the chronological telling of life's history features section previews that introduce visitors to the next geologic period (Fig. 7.4).
- *Theme Labels.* Within an individual display, a theme label carries the main messages of that display. These labels are written with both strollers and studiers in mind (see Chapter Four).

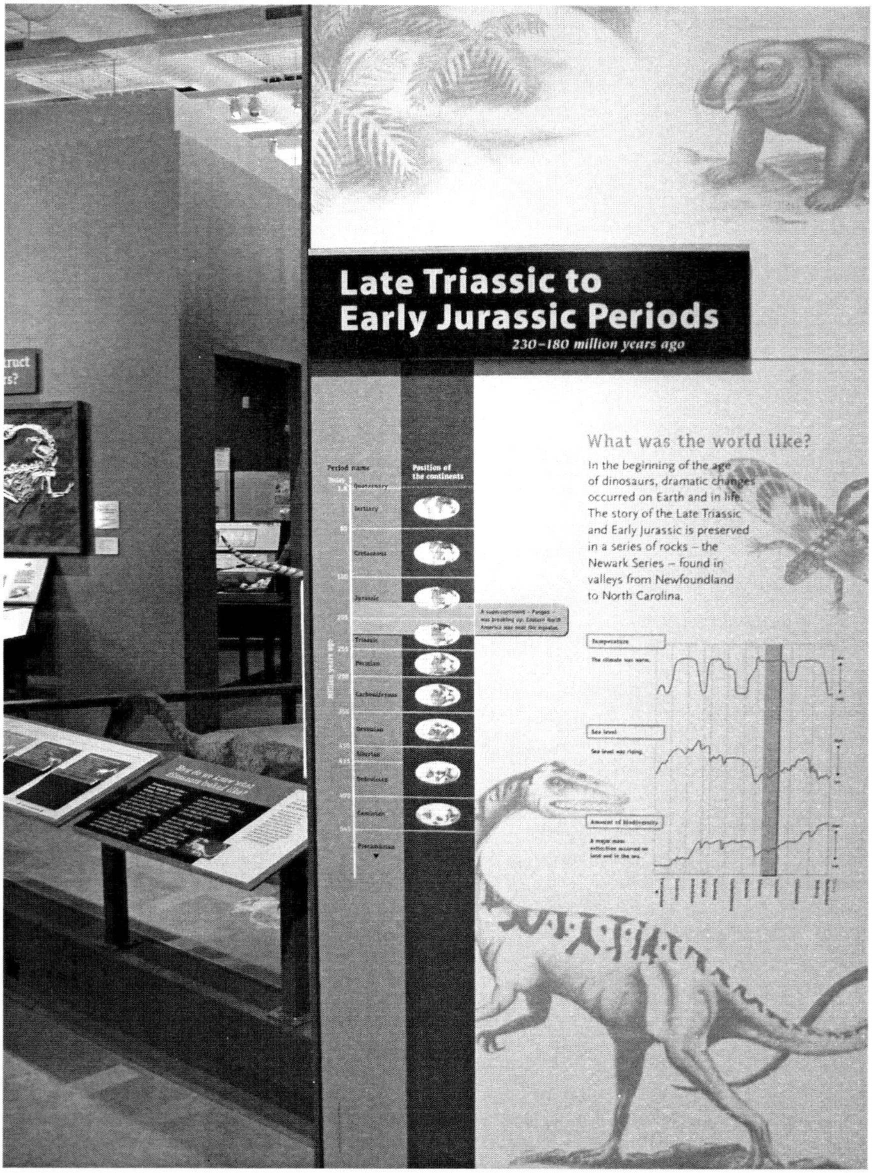

FIGURE 7.4. A chronologic story through Earth's history, the permanent galleries at Museum of the Earth feature distinct section previews to orient visitors along their journey and introduce each section. Photo by R. Kissel.

- *Focus Labels.* One or more focus labels present additional in-depth information beyond that of the theme label. Focus labels are written with studiers in mind.

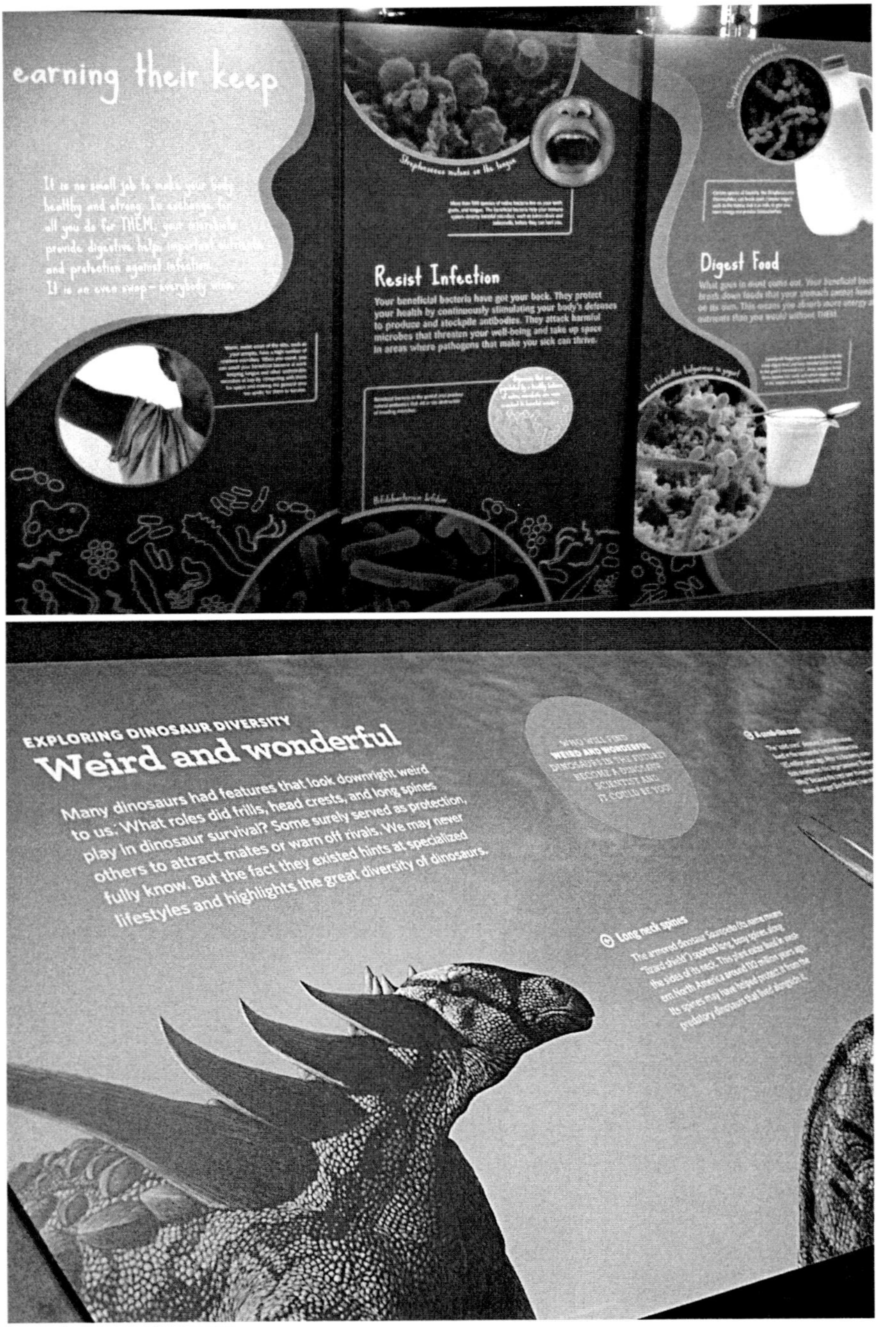

FIGURE 7.5. In these displays from the Discovery Place (Charlotte, NC; top) and the Natural History Museum of Los Angeles County, the hierarchy and relationship of the label text is evident from the physical layout and graphic design. Photos by R. Kissel.

Theme and Focus labels coincide with the Primary and Secondary interpretive labels, respectively, of Rawson (2006), and the Group and Detail labels of Lord and Lord (2001). Well employed, the use of theme and focus labels, as well as captions, within a single display is an effective strategy for reaching a variety of visitors. When considering displays within an exhibition, a team must ensure that all elements—from writing to design—are complementary and the hierarchy is intuitive to the visitor (Fig. 7.5). A perfectly crafted label is worth little if its design and placement don't support the label's content and message. For any successful exhibition, writing and design are intimately linked, and communication among members of a team is required to produce that necessary harmony.

Finally, there will always be those content pieces that are so very interesting but not necessary for the narrative. If an exhibition team finds such information interesting, then it's likely that visitors will too! Feel free to keep these small nuggets and classify them as Fun Facts. While not part of the narrative directly, Fun Facts can add color to any exhibition, as long as the tone of the exhibition warrants their inclusion. Fun Facts should possess a distinct and repeated graphic design, immediately telling the visitor that their content is supplementary information to the primary storyline (Fig. 7.6).

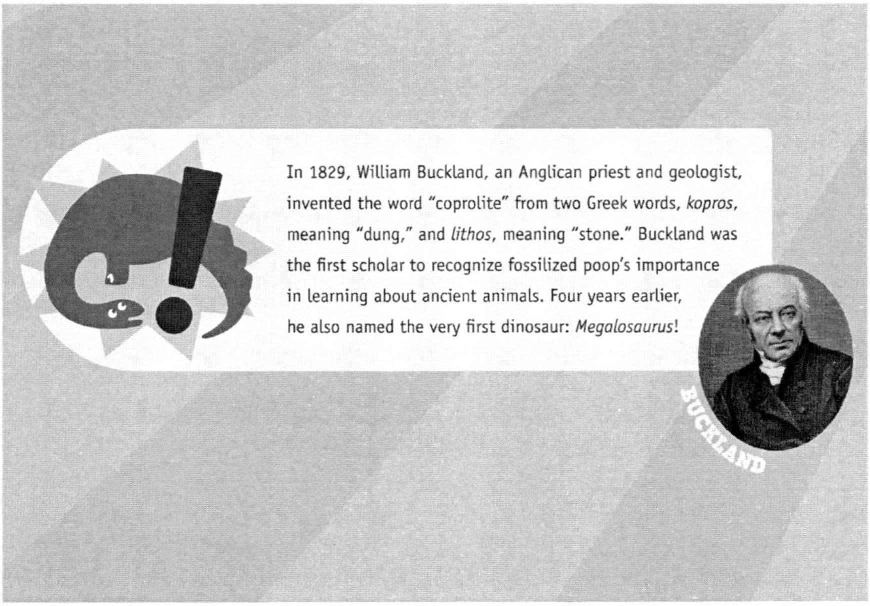

FIGURE 7.6. A Fun Fact from Museum of the Earth's temporary exhibition *Did Dinosaurs Poop?* provides information that, while not necessary for the storyline, adds information that visitors might find interesting. Courtesy Beth Stricker, Paleontological Research Institution.

Label Writing 101: Initial Considerations

What makes a good label? The very best labels are engaging, entertaining, and educational; they encourage action, conversation, and reflection. They consider all visitors, from the professional/hobbyist to those adults acting as facilitators for younger audiences. While such adjectives and ambitions are delightful to consider, they describe a product, not a process. The drafting of any good label begins long before any visitor enters the exhibition on opening day.

Once the development of any exhibition begins, one of the very first questions to ask is: What does the visitor already know—or think they know—about the topic? Front-end evaluation is important (as discussed in Chapter Four), and it can certainly drive the level of depth ultimately presented within the exhibition. A front-end evaluation on a new glacier display within Ithaca's Museum of the Earth produced surprising results, for example. Despite the fact that the local landscape was heavily influenced by the glaciers of 20,000 years ago, producing the area's famous gorges and waterfalls, the evaluation showed that visitors to the museum had only vague concepts of glaciers (Cronin et al., 2012). The results forced the development team to step back and reconsider approaches; the evaluation completely informed the process. And evaluation should continue through the writing process, by using formative evaluation to test the effectiveness of the label writing.

In addition to establishing visitors' prior understanding of a topic, another consideration is voice and tone. Voice answers the question: Who is speaking to the visitor—an expert in the field, such as a curator or research scientist; an enthusiastic tour guide; or perhaps a juvenile *Stegosaurus* from 150 million years ago? Consideration of both the exhibition's topic and the intended audience will help determine the proper voice and tone. Exhibitions developed for younger audiences, for example, will often carry a more playful tone. Developed by the New York Hall of Science, *Charlie and Kiwi's Evolutionary Adventure* is written and designed to relay concepts of adaptation and natural selection to school age children. Key to its success is that the story contained within the exhibition revolves around a young boy named Charlie, who is writing a report on his favorite bird, the kiwi. Charlie provides the voice and sets the tone for not only the writing but also a playful graphic design (Fig. 7.7); the use of characters can provide an engaging, accessible solution to the presentation of a traditionally academic topic (Palmquist 2001). And while the labels of the exhibition are targeted and therefore accessible to *younger* audiences—five- to twelve-year olds, specifically—it is important to note that they are engaging for *all* audiences. While an accessible label might, unfortunately, not be engaging, every engaging label must first be accessible.

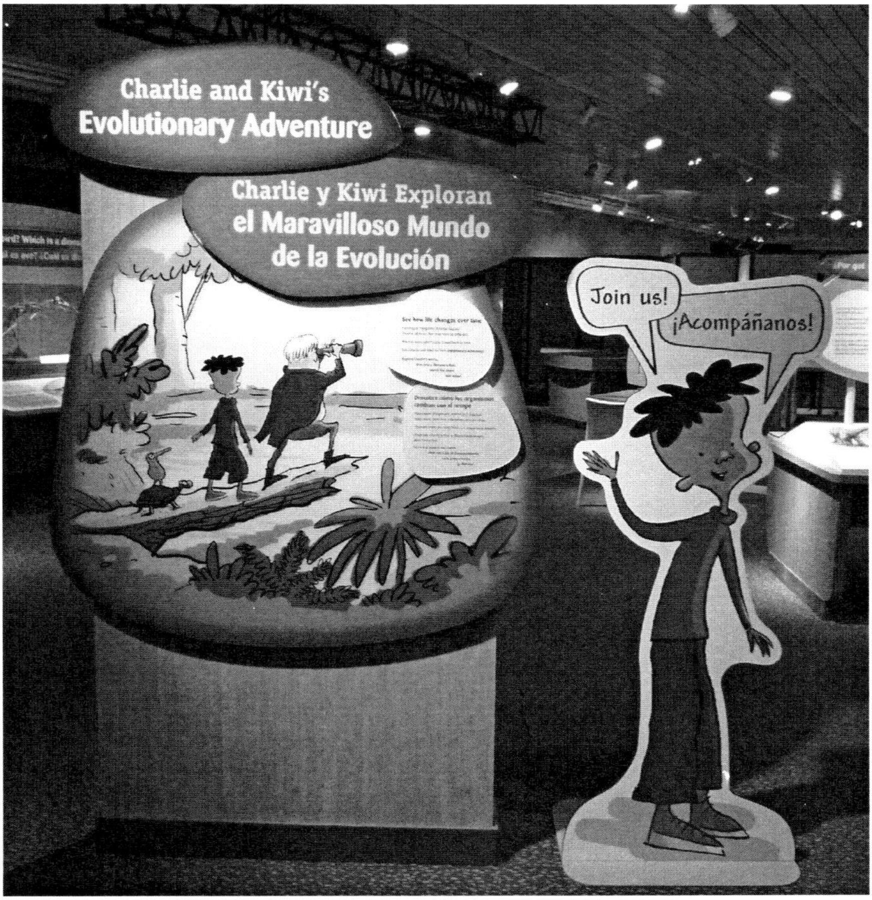

FIGURE 7.7. The playful tone of the story told within *Charlie and Kiwi's Evolutionary Adventure* is reflected in the graphic design. Jeff Kennedy Associates.

In contrast to the voice and tone of *Charlie and Kiwi's Evolutionary Adventure,* many of the permanent halls at the American Museum of Natural History possess a highly academic, curator-driven voice, as in the following example:

Ostracoderms, the jawless first vertebrates

These primitive vertebrates were armored by bones in the skin, a feature that gave rise to the name ostracoderm ("shield skin"). Our vertebrate ancestors lived in the water, not on land. Primitive vertebrate features are therefore adaptations to an aquatic lifestyle, and are related to improved locomotion, feeding, awareness, and protection. Ostracoderms lacked jaws, and probably filtered food particles from water that entered the mouth. The only surviving jawless vertebrates are the lampreys and hagfish.

In these scenarios, the museum is the teacher and the visitor the student; facts are provided for the learner, and while the labels might be praised for carrying a certain level of authority, there is an established intellectual divide between the writer and the reader, whether intended or not. The visitor is not a part of the scientific process; he or she is told about it. The fossil halls at the Los Angeles Natural History Museum offer a different voice. There, the voice—while still confident and authoritative—is less academic and far more engaging for the visitor, asking questions to prompt discovery and the visitor's participation in the story:

What were they like?

Today we know dinosaurs only from fossils. But these old bones and bits of skin once belonged to living, breathing animals. They hunted for food and fled from danger. They sought mates and cared for their young. They suffered diseases and injuries. They faced life's challenges just as we do today.

These examples both introduce major animal groups, and they demonstrate how the presentation of similar topics—intended for similar audiences—can carry a different voice and tone. The voice should match and embrace the personality of the institution,[2] and the best interpretive labels present an appealing voice that is not preachy or pedantic, but also not too simplistic or condescending (Serrell 1996). Even the images and words of children, used effectively, are capable of the delivery of science-based concepts (Fig. 7.8).

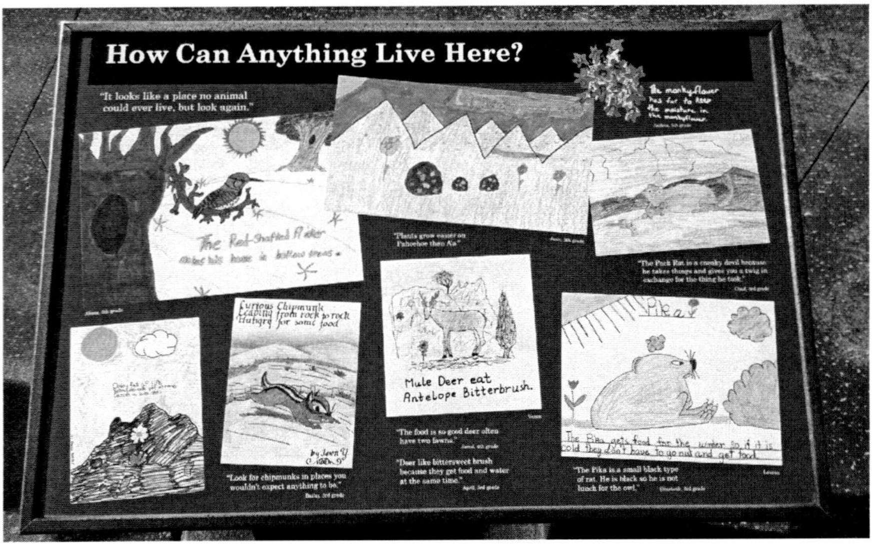

FIGURE 7.8. Outdoor signage at Craters of the Moon National Monument and Preserve effectively presents the local ecosystem with the use of children's art and words. Photo by R. Kissel.

With all this talk about label hierarchy, we don't want you to forget the established Big Idea and stated goals. Remember, the Big Idea is the primary message of the exhibition. For the visitor, it is *the* take-home message; for the development team, it is the guiding message that informs all decisions. Write every label with the Big Idea and learning goals in mind; straying from the Big Idea can only lead to an unfocused narrative and confusing messages for the visitor. Check every label against the Big Idea. The established learning goals can also help prioritize messages. During the development of every exhibition, the problem of *too many* ideas is always a constant, ultimately leading to the question: Which ideas do we include? Defined learning goals greatly assist with that prioritization.

Finally, before the first word is written, remember that writing takes time. An ideal allotment of time is four hours per label, not including review and revisions. Imagine an exhibition with 200 labels; that's 800 hours, or 20 full-time weeks. Having adequate time built into the development schedule ensures more opportunity for review and revision, providing the chance to perfect the language, tone, consistency in message, and consistency in presentation (Fig. 7.9). And writing ideally occurs before detailed designs are produced. The story of an exhibition—as told by the objects and labels, including content and tone—informs both the 2D (graphics) and 3D design. The curatorial voice of the American Museum's fossil halls, for example, is in harmony with the simple look and feel and the classical architecture of the halls, while the voice of the fossil halls in Los Angeles—that of a friendly teacher—is matched with a physical space featuring bright murals and a host of bold colors.

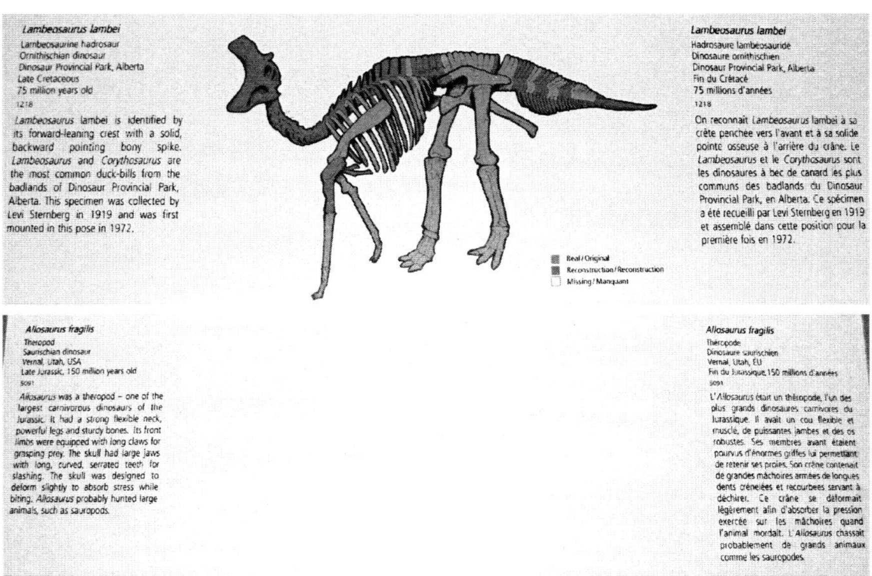

FIGURE 7.9. Within the *James and Louise Temerty Galleries of the Age of Dinosaurs* at the Royal Ontario Museum, object labels and captions are unfortunately inconsistent, with some (top) indicating which bones are real, while others (bottom) lack this information. Photos by R. Kissel.

Label Writing 101: Pencil to Paper

After evaluation and content research, it's time to write. As the pencil hits the paper, the possibilities are endless—for better *and* worse. Focus is key. And while there is no exacting formula that without fail will produce an engaging, memorable label, there are fundamental guidelines that can lead you along the way.

- Write to the action, allowing the message to emerge from that action. When presented with the label in the exhibition, what will visitors be seeing, doing, or touching? Leading with this action in mind creates a stronger connection between the label and not only the object but also the visitor. It incorporates the visitors into the exhibition, as demonstrated by the following award-winning label (Russick and Nordstrom 2010) on beluga whales at Chicago's Shedd Aquarium:

  By spitting, belugas uncover crabs to eat

  Watch for belugas spitting streams of water. They're not just playing! In the wild, belugas explore the seafloor, spitting to clear away sand and silt. That's how they find crabs, worms and octopuses to eat.

  Invitation is an effective way to immediately engage the visitor with the exhibition, and it is especially effective within a label's heading.

- Regardless of the topic's complexity, remember that a single label should carry only one idea. In this way, each label is a small story. And stories are specific. Taken together, they add up to the larger narrative that is the exhibition. Within The Field Museum's dinosaur hall, the specimens are grouped based on their evolutionary relationships, and within these displays are stories of diet, lifestyle, or size:

  This giant was a middleweight

  At 72 feet long and weighing 33 tons, *Apatosaurus* was neither as large as the largest sauropod (115 feet long), nor as small as the smallest (20 feet long). Still, *Apatosaurus* was massive.

  Some whales we know today rival sauropods for size, but no land animal before or since has ever come close.

  Collectively, labels such as this one reinforce the hall's theme of dinosaurs' tremendous diversity—both physically and taxonomically. The label also contains what Serrell (1996, 90) calls a "snappy ending," which provides readers—in the form of closure, reinforcement, or new insight—with "a reward for reading it all."

- The information found within the interpretive labels of an exhibition is best served with a predetermined and consistent organization. For theme and focus labels, headings are recommended. As shown in the two preceding examples, the headings (in bold) provide a point of engagement and entry for the visitor, and prepare the visitor for what he or she is about to read. Captions may carry headings, but they are not necessary, as the specimen, artifact, or image provides the initial point of engagement. The body of the label should consist of short paragraphs, ensuring that the information is presented in easily digestible portions.

- Ask questions! Another technique for engaging visitors is to ask questions. In any exhibition, asking questions disrupts the potentially overwhelming presentation of static content, providing a much-needed breather. But they also serve a greater purpose. Questions in labels may help the visitor become part of the exhibition's storyline, or help inspire curiosity and wonder about a particular topic (Fig. 7.10). They are not intended to stump the visitor, and it is critical that—even if it is not provided within the label, the visitor is still capable of coming up with the answer. Maintaining a visitor's comfort ensures a satisfactory museum-going experience; asking questions that assume prior knowledge or require guesswork (Fig. 7.11) is potentially frustrating. For science-based exhibitions, the inclusion of questions serves an additional purpose. Whether the discipline is anthropology or zoology, all scientific investigations themselves begin with a question (Lederman 2009). Thus, by simply including labels in which the content is prompted by a question, the process, methods, and characteristics of science—the Nature of Science—is embedded within the exhibition. Presented properly, questions encourage inquiry, observation, and discussion. Finally, it is important to carefully consider the intent and consequence of any question. For example, labels beginning with "Did you know?" are often more correctly asking, "Guess what *we* know?" This technique is commonly used to present interesting but obscure content-related facts that few to no visitors will know. As a result, these questions are unintentionally reinforcing visitors' lack of knowledge; they are exclusive.

- The text found within labels is ideally conversational; write for reading aloud. Beginning sentences with "and," "but," or "or" is not only acceptable but recommended to produce emphasis or provide a logical transition. Consider the following examples, which focus on Richard Owen's rejection of Darwin's ideas:

  — Following the publication of Origin in 1859, many of Darwin's colleagues quickly embraced his theory, but not all. Prominent zoologist Richard Owen...

  — Following the publication of Origin in 1859, many of Darwin's colleagues quickly embraced his theory. But not all. Prominent zoologist Richard Owen...

Note the shift in emphasis. In the latter example, the reader is more adequately prepared for the content that follows—Richard Owen as a vehement opponent of Darwin's ideas—and he or she is informed of Owen's position even before reading his name. Also related to conversation, engage the visitor by addressing him or her. Using "you" can transform written text—a monologue—into a much more inviting dialogue.[3] Ravelli (2006, chapter two) provides a wonderful overview on the importance of language and its organizing framework within museum labels.

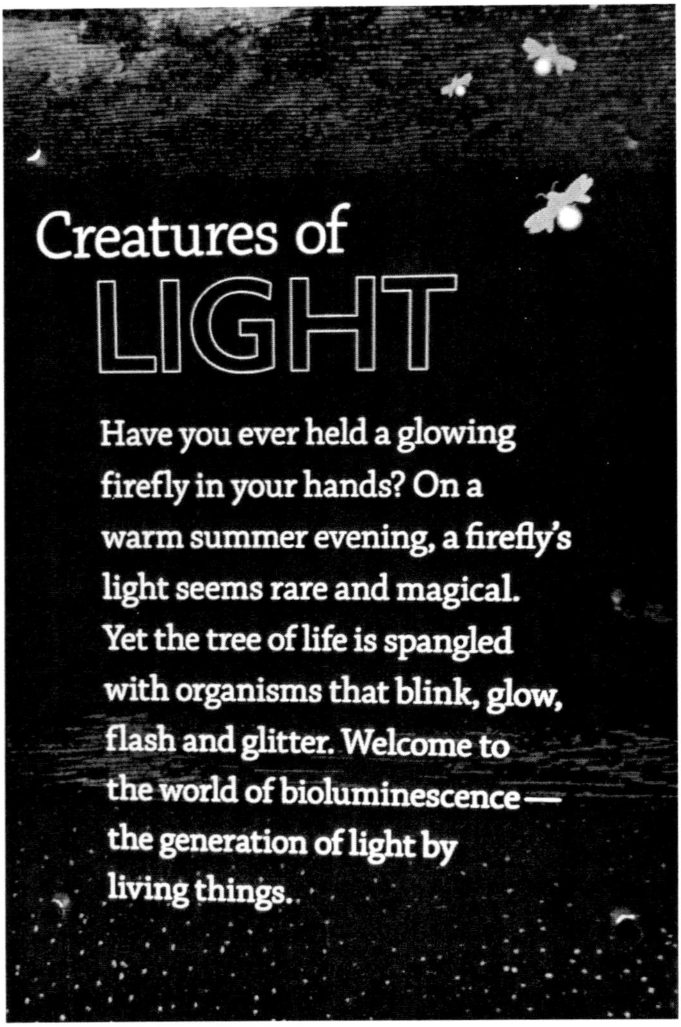

FIGURE 7.10. With a simple question, this introductory label from the *Creatures of Light: Nature's Bioluminescence* temporary exhibition at the American Museum of Natural History brings a visitor's own experiences into the exhibition's narrative. The label also effectively defines a potentially tongue-twisting scientific term. Photo by R. Kissel.

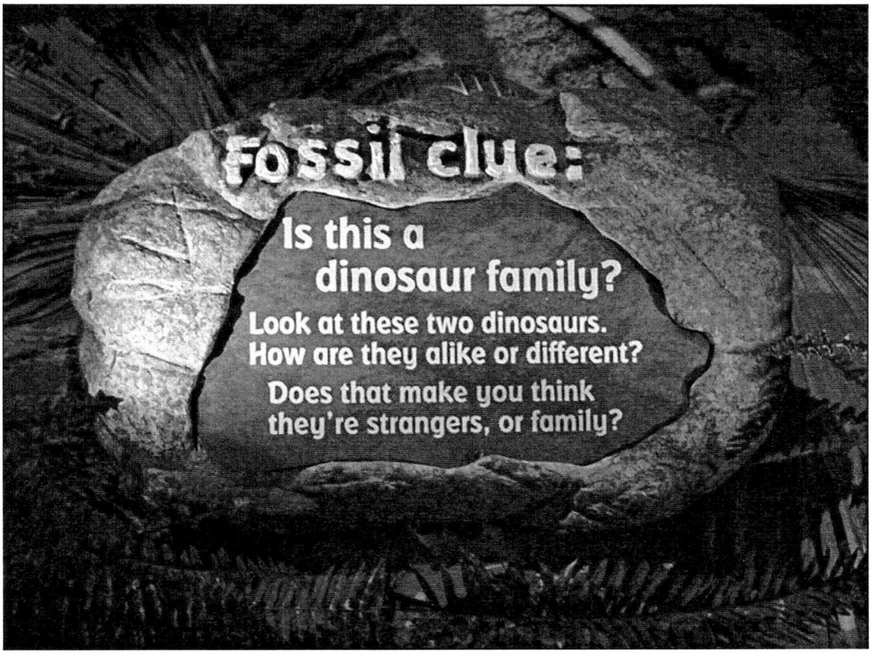

FIGURE 7.11. At the *Dinosphere: Now You're in Their World* exhibition at The Children's Museum of Indianapolis, visitors are asked to make scientific conclusions without enough information to allow a confident—and therefore comfortable—ruling. Photo by R. Kissel.

- Make the content relevant. The concepts and principles of many scientific disciplines, while second nature to curators and other specialists, might be abstract to many visitors. It is therefore important to help the visitor grasp the importance, uniqueness, or scale—large or small—of the information. In the following example from The Field Museum's *Evolving Planet*, the serpentine tail of the dinosaur *Apatosaurus* is discussed.

A terrific tail

Apatosaurus might have used its tail, which ended in a series of small rod-shaped bones, as a mighty whip to defend itself against threatening predators. Some sauropod tails had up to 80 individual vertebrae. (You have 33 in your entire backbone.)

Without the final sentence, the impressive number of tail vertebrae found within some sauropods would be meaningless to many visitors. But by comparing it to a familiar entity—the visitor's own body—the presence of 80 vertebrae possesses considerably more weight and has more meaning. A similar example is found within the discussion of the difference between weather and climate. The term climate refers to the prevailing conditions of a region, including temperature, humidity, precipitation, etc., throughout the year, averaged over a series of years. Weather, on the other hand, refers to the state of the atmosphere with respect to temperature, wind, cloudiness, etc. at a particular place at any given time. Providing these or similarly academic definitions in an exhibition is acceptable, but they are significantly enhanced if followed by: *Weather tells you how to dress in the morning; climate tells you what to have in your closet.* By including this short statement, the scientific concepts of weather and climate are relatable to visitors' lives and everyday experiences. They are more meaningful and, no doubt, visitors will have a heightened rate of retention.

- With science-based exhibitions, never be afraid to visit concepts that might seem too basic at first. A perfect example is the term *theory*—a term very much misused, even in many academic writings (e.g., Montgomery 2012). In typical conversation, the term is used as a replacement for *guess* or *hunch*. But in science, the term theory has a very specific meaning: a theory is an evidence-based explanation of a naturally occurring phenomenon. The theory of evolution by means of natural selection, for example, is not mere speculation but instead an idea supported by more than 100 years of rigorous testing and research. Defining such terms at the onset of any science-based exhibition will properly prepare visitors for their journey through the exhibition (Fig. 2.1). Of course, a vocabulary that is beyond many visitors is best avoided when possible, but it is sometimes necessary (Fig. 7.10); embrace it as a product of science if its inclusion can increase understanding of the topic and, more importantly, increase scientific literacy beyond the museum visit.

- Be creative! Inspire visitors' curiosity with unexpected statements or unusual viewpoints; surprise and intrigue the visitor.[4] While presenting information in a straightforward manner might be efficient, it is matter-of-fact and likely forgettable. To state that *Tyrannosaurus* and *Stegosaurus* were separated by 85 million years, for example, is far less engaging than explaining that *Tyrannosaurus* lived closer in time to us today than it did to *Stegosaurus.* In the latter presentation, the content is uniquely presented and relatable, bringing the visitor into the story of what is otherwise a list of scientific names and dates. It also takes advantage of prior understanding of *Tyrannosaurus* as a beast from long, long ago, using that notion to build a deeper appreciation of the extent of dinosaurs' long reign. A small creative flourish can lead to a much more memorable label (Figs. 7.12 and 7.13).

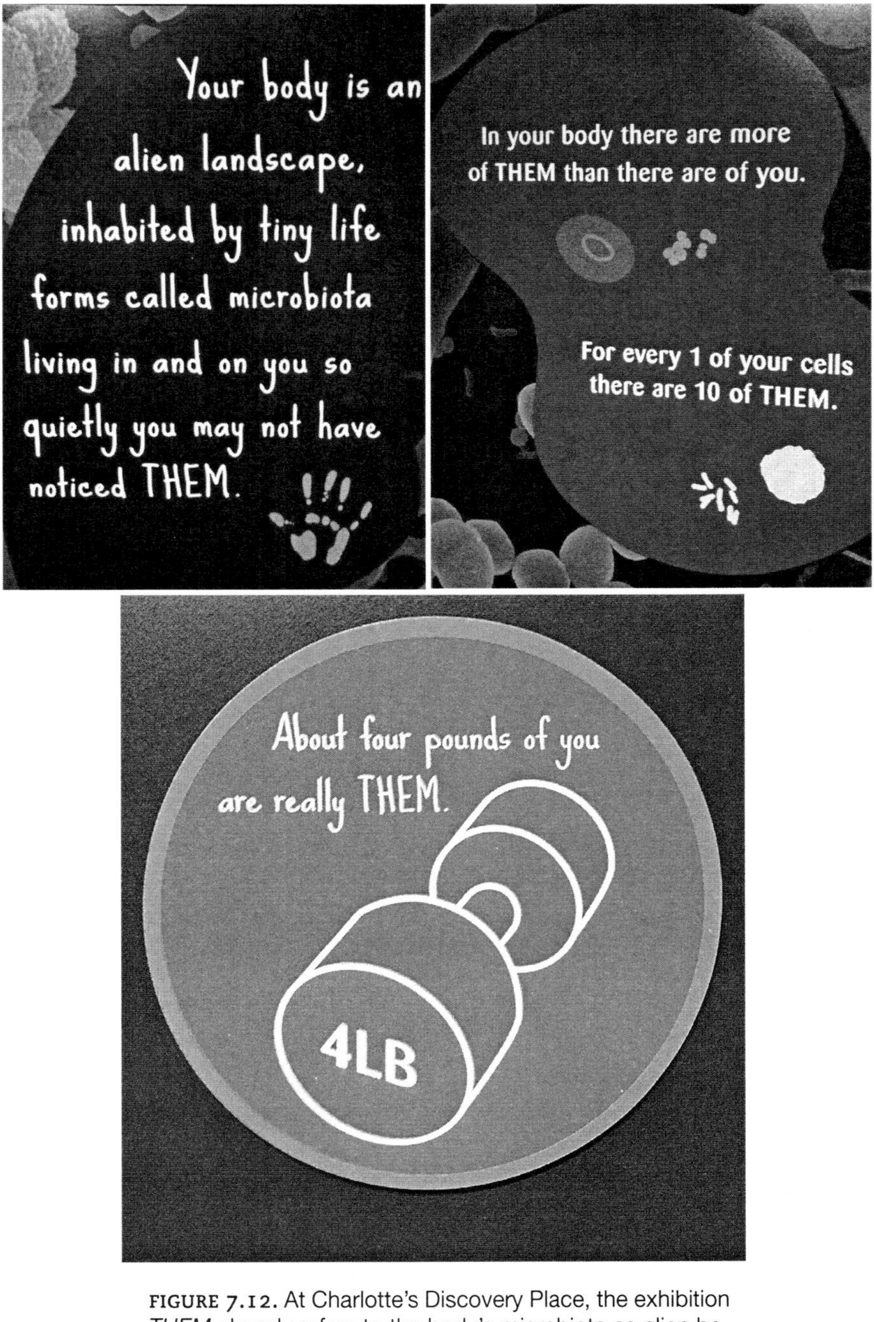

FIGURE 7.12. At Charlotte's Discovery Place, the exhibition *THEM* cleverly refers to the body's microbiota as alien beings, or "THEM." Photos by R. Kissel.

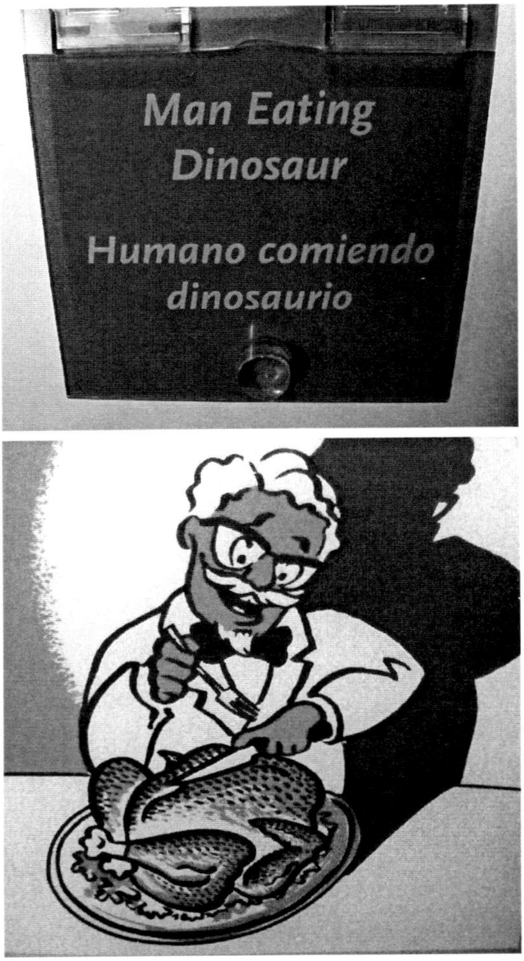

FIGURE 7.13. A flip label at the New Mexico Museum of Natural History and Science plays on the concept of dinosaurs as bloodthirsty savages to reveal a simple but beautiful reinforcement of birds' true evolutionary origin. Photos by R. Kissel.

A final consideration is the use of bilingual labels. Garcia-Luis et al. (2011) reported that only 10 percent of institutions in the United States offer most or all visitor information in more than one language, while 55 percent of examined international institutions offered most or all information in more than one language. When not given the attention they deserve, bilingual labels can result in an awkward and often befuddling experience. But when done effectively, they open up exhibition content to more visitors, and so their inclusion should be a serious discussion within any exhibition team. If bilingual labels are pursued, Plaza (2009) provides the following guidelines:

- Both languages are equally important. Equal care and editorial emphasis should be given to labels in both languages, and any exhibition schedule should consider this dual importance and equal time.

- Translation requires re-contextualization. It is important to translate the content correctly, yes, but it is equally important that the text is reinterpreted and edited to maintain a consistent meaning and voice across languages, despite variations in idiomatic and colloquial expressions.

In addition to presenting accurate content, there are significant design considerations when dealing with bilingual labels, and these need to be considered from the outset. If both languages are equally important, as recommended in one of the preceding bullets, then both languages should be given equal square-footage and visual weight, while graphically maintained as distinct delivery methods (Fig. 7.14).

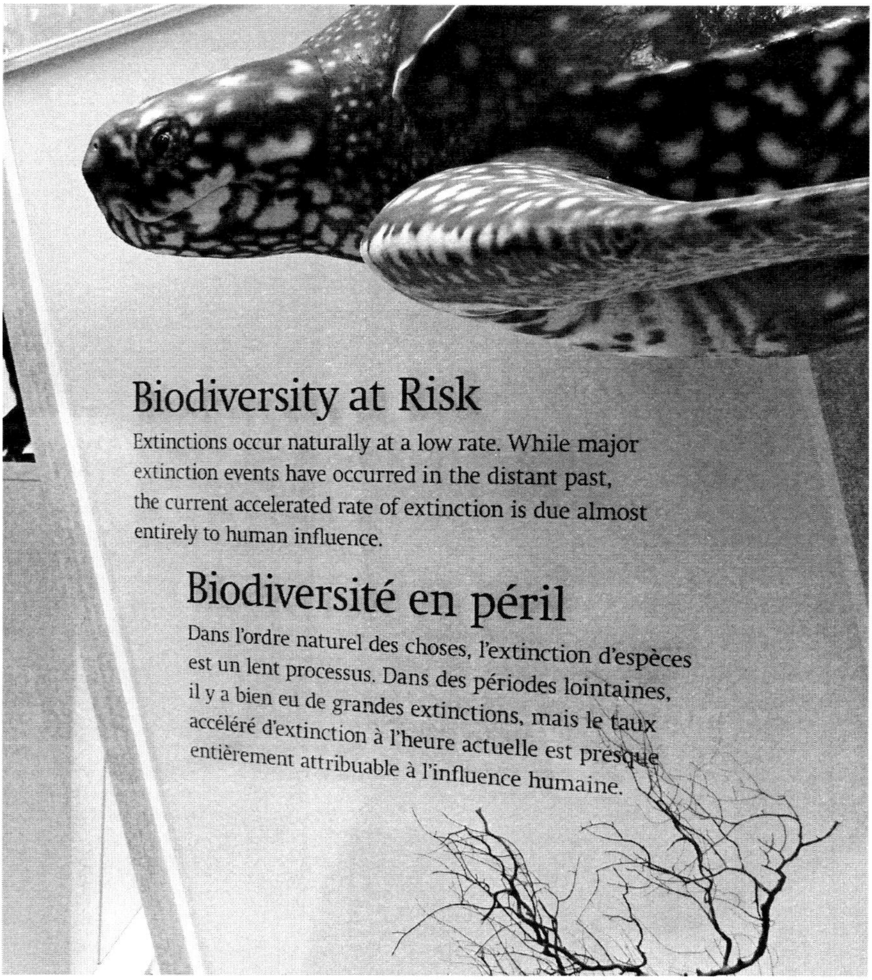

FIGURE 7.14. Wonderful example of a bilingual label featured in *Life in Crisis: Schad Gallery of Biodiversity* at the Royal Ontario Museum. Photo by R. Kissel.

A Brief Note on the Myth of Dumbing Down

One of the greatest myths concerning science-based exhibitions—and science education in general—is the need to "dumb down" the content. It is a phrase that arises time and again during the development of many exhibitions, and it's unfortunate. As educators—and every member of an exhibition team is an educator—we must respect our audiences. The moment that a team member considers the visitor "dumb" is the moment that he or she should leave the team. No one individual is ever the sole authority on any given topic, let alone the many topics, found within natural history museums, zoos, and aquaria, so it is the responsibility of the exhibition team to translate objects and the ideas and interpretation that they generate into a story that is accessible and satisfying to all visitors—from experience seekers to hobbyists. The moment that one uses the phrase "dumbing down" is the moment that he or she admits failure as a communicator and educator. Any good scientist, educator, or exhibition developer can take any concept—no matter how academic—and present it in a manner that any non-specialist will understand and appreciate. It is the exhibition developer's challenge; it is the exhibition developer's skill; it is the exhibition developer's duty. Value the visitor, and value your responsibility to them.

Conclusion

Within any exhibition, space is always limited. Coupled with the knowledge that labels exceeding 150–200 words are rarely read (Hein and Alexander 1998), and while Lord and Lord (2001) suggest that visitors stop reading labels after the first or second paragraph, the most important concept of label writing is clear: every word matters. Much consideration is required before the very first word is written. Who is the intended audience? What is the voice and tone? And what is the perfect balance between authoritatively presenting science and content and achieving accessibility? If the voice is too academic and the resulting labels too difficult to follow, the writing has failed. It is critical that the writing—including its voice, tone, and intended audience—inform the design. The visual presentation of the content must match the written presentation, and if the visitor is required to encounter a series of labels in a sequential manner, for example, the design needs to support that intended sequence. It can never be overstated that all elements of an exhibition are complementary.

Material of the Visual

"It's not what you look at that matters, it's what you see."

—Henry David Thoreau

There are many ways to communicate with your audience. Labels, discussed in Chapter Seven, spring to mind as one of the most obvious. But visual clues beyond labels might, in many cases, communicate as profoundly as a complete sentence. A lot about a given topic can be conveyed without the use of words; images, color, lighting, ambient sound, temperature, and smell all work together to create a visitor experience beyond engagement with didactic text panels. Both 2D and 3D elements combine with an exhibition's content to create an experience, and it is the 2D material of the visual that is examined in this chapter.

We begin by exploring some of the basic principles of exhibition graphics and text panels, including typographic design, panel layout, and images—covering both permissions and the preparation of digital files. We then look at some of the tools of the trade as they relate to an exhibition's overall look and feel, including color, materials, and finishes.

Typography

The previous chapter's discussion of writing label copy segues nicely into a conversation about the visual nature of this text. Typography is the art or process of printing with type. Whether or not a visitor engages with an exhibition's labels is determined not only by the words presented but in large part by its typographic design and legibility. Legibility—the ease with which a person can recognize and understand words and symbols—is influenced in large part by familiarity; at its bare bones, it is the idea that we recognize what we know (Serrell 1996). And this in turn impacts readability, or "the ease of reading" (Majewski, n.d., 3). What follows is an ode to legibility, and ultimately readability, that focuses on typeface, text layout, and size. These recommendations greatly increase the potential engagement of an exhibition's labels.

Typeface

Though it might seem a bit esoteric at first, it is important to have at least a nominal understanding of typographic terminology in order muddle your way through practical applications. Font is a common word, and in today's world it is often used interchangeably with the term typeface. The designation between the two was particularly important before the advent of digital publishing when printed text was set using individually cast letters, but in today's parlance the terms are essentially interchangeable, and whether you use *font* or *typeface,* most will know what you are talking about.

A typeface is broken down into a number of component parts, including ascenders, x-height, and descenders (Fig. 8.1). The x-height—defined as the height of the lowercase letter x—is the height of a particular typeface's lower case letters (not including the ascender and descender letters). The x-height is an important consideration because it affects "the feel of a typeface, how many characters fit on a line, and—depending on how the type is set—how easily...text can be read" (Adobe 2000, 3). When taken collectively, a typeface's ascender + x-height + descender = its point size. This has very real practical implications when you consider that a typeface with a small x-height relative to its ascenders and descenders will look much smaller than other typefaces of the same font size (Fig. 8.2).

The idea of 12-point font is probably not new (it is a classic requirement of many college term papers), but what is not as commonly known is its actual physical measurement. One point is equal to 1/72 of an inch, making 12-point font, for example, 1/6 of an inch. Not surprisingly, font size has a significant impact on legibility. Have you ever been to an exhibition where you had to squint to read the label? Or you have you ever come across a panel of exhibition text that looks like a manifesto and you quickly made your way past it, rather than stopping to read all of that tiny print? If you keep a few general guidelines in mind, you will be able to avoid these scenarios with your visitors.

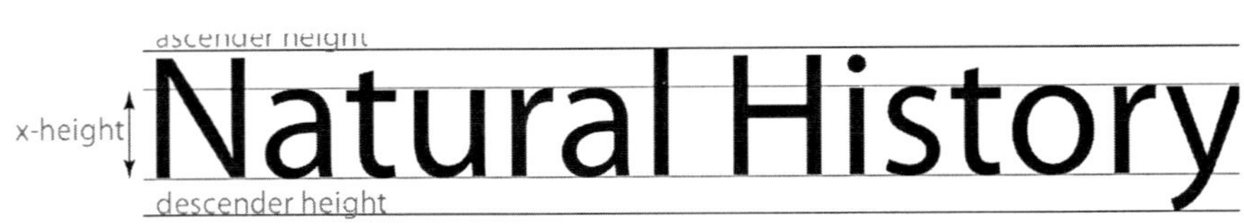

FIGURE 8.1. The typeface of Natural History includes its ascenders, x-height, and descenders; together, they make up a typeface's overall point size.

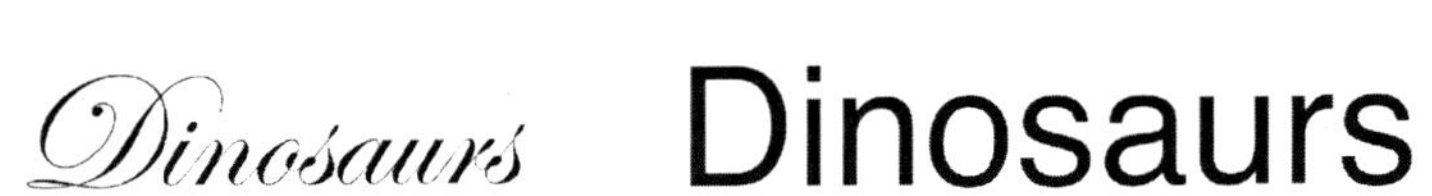

FIGURE 8.2. The word dinosaurs written in the same font size in Edwardian Script (left) and Helvetica (right).

While the general readability of a typeface is impacted by font size, reading distance is perhaps the single most influencing factor related to ideal size. How far away should a visitor be in order to comfortably read a particular graphic? As a general rule, exhibition text should be no smaller than 18-point font. Table 8.1 provides the guidelines suggested by both the Smithsonian Institution and Parks Canada with regard to font size and expected viewing distance. Thinking about font size relative to the general label hierarchy, as discussed in Chapter Seven, is also useful. See Table 8.2 for guidelines on this topic.

TABLE 8.1. General guidelines relating font size and expected viewing distance

Probable Viewing Distance	*Minimum Type size (x-height)*	*Set size (point)*
Less than 75 mm (3 in)	4.5 (3/16)	24
1 m (39 in)	9 (3/8)	48
2 m (78 in)	19 (3/4)	100
3 m (118 in)	28 (1-1/8)	148

Source: Majewski's (n.d., 24) Smithsonian Guidelines for Accessible Exhibition Design

TABLE 8.2. General guidelines relating font size and label hierarchy

Exhibit Elements	*Type Size*
Main Titles (Identifiers)	To be sized as appropriate for the size of the exhibition unit, and the context of viewing. 2-3 inches (5-7.5 cm) often used as the standard
Subtitles (key points)	100-120 point
Theme Label text (explanatory text)	30-40 point
Focus Label text (discussion)	24-32 point
Caption and Object Identification Label text (individual/group labels)	18-22 point

Source: Based on Spencer (2001, 398-399)

In addition to font size, consideration of lighting conditions and visitor numbers is necessary. Is there glare or shadow, for example, and will crowds impact how close a potential reader can get to the text? The more environmental impediments to the visitor seeing the type well, the more the size and contrast of the type should be increased in order to compensate.

A Tale of Two Serifs

As shown in Figure 8.2, two different fonts of the same point size can appear dramatically different and, as a result, have remarkably different legibility. So, with the seemingly endless supply of font styles available, where does one begin? How do designers pick which fonts to use? If you stick with the basics, you can get by with two main kinds of typefaces in your typographic tool kit: the serif and the sans serif. These typefaces are generally the easiest for the widest audience, including those with low vision, language problems, and cognitive disabilities (Majewski n.d., 19). Serif fonts include a short brushstroke, cross line, or little kickstand at the end of each stroke. Some scholars have suggested that this practice originated with the Romans and was a result of painting letters onto stone slabs before they were cut (Catich 1968). As a result, serif fonts are often associated with classical or traditional looks (Serrell 1996). Sans serif fonts (French for *without serifs*), on the other hand, lack the characteristic crosshatch of their serif cousins. These fonts can be described as "clean" and "modern" (Serrell 1996, 194).

Generally speaking, serif text fonts are easier to read than sans serif fonts and are often used for body copy (Adobe 2000, 10). When considering typeface for titles and subheads, there is more room for variety than recommended for body text, as they are considered more visual in nature and are intended to grab a visitor's attention, not necessarily explain content in depth. But do keep in mind that no matter where its location, text set in all capitals can be significantly harder to read than mixed text (Adobe 2000, 11).

When considering fonts, it is best to avoid light, condensed, or extended typefaces, as well as script and italics, as these are often difficult for visitors with low vision (Majewski n.d., 21). As with everything, there are always exceptions to the rule, and a designer should continuously consider the overall tone and the look and feel of the exhibition, even in relation to font choices and how they relate to the exhibition's Big Idea and goals. When creating a science-based exhibition, for example, certain rules dictate the use of italics; in these instances (see Chapter Seven), content rules and special typeface forms like italics should be used.

Layout

The months and months spent drafting exhibition copy confirm—without hesitation—that exhibition text is written with the intent and hope that visitors will read it. But because of the nature of the context, even the most beautifully written label copy competes with the host of factors that are the visitor experience—different from the focused attention you are giving the written text within this book. If the goal is to maximize the legibility and ultimately the readability of label copy, aspects in addition to general typeface design—serif versus sans serif—and font size must be considered. The design layout of label copy also impacts legibility and ultimately readability. Here are a few guidelines to keep in mind when laying out text.

- Use aligned left and ragged right margins.
  - Not centered—"a predictable beginning point, line after line, and evenly spaced words are much easier to read..." (Majewski n.d., 24).
  - Not ragged left margins—these make it harder to find your way back to the beginning of the next line in a paragraph (Serrell 1996).
  - Not justified—justified text runs the risk of incorporating large swaths of empty space, or rivers, that create uneven gaps between words. The text can start to look like a visual block, in which case the eye is primarily drawn to the shape of the text block instead of the words within the block, causing the reader lose the trees for the forest.
- Avoid word divisions that require a hyphen at the end of a line.
- Avoid widows and orphans. This isn't the heartless statement it first appears to be! Widows and orphans are used to describe words or short lines that fall at the beginning or end of a paragraph and are left dangling at the top or the bottom of the page. According to *The Chicago Manual of Style*, a widow is "a paragraph-ending line that falls at the beginning of the following page/column, thus separated from the rest of the text." And an orphan is "a word, part of a word, or very short line that appears by itself at the end of a paragraph. Orphans result in too much white space between paragraphs or at the bottom of a page."
- Certain line lengths facilitate reading. Too many characters complicate the process, so instead of trying to fit as many words as you can on each line, aim for an average of 50 characters (Majewski n.d., 17; McLean 1993, 113).

Figure 8.3 illustrates a series of labels—from the fossil halls at the American Museum of Natural History—that break many of the preceding rules.

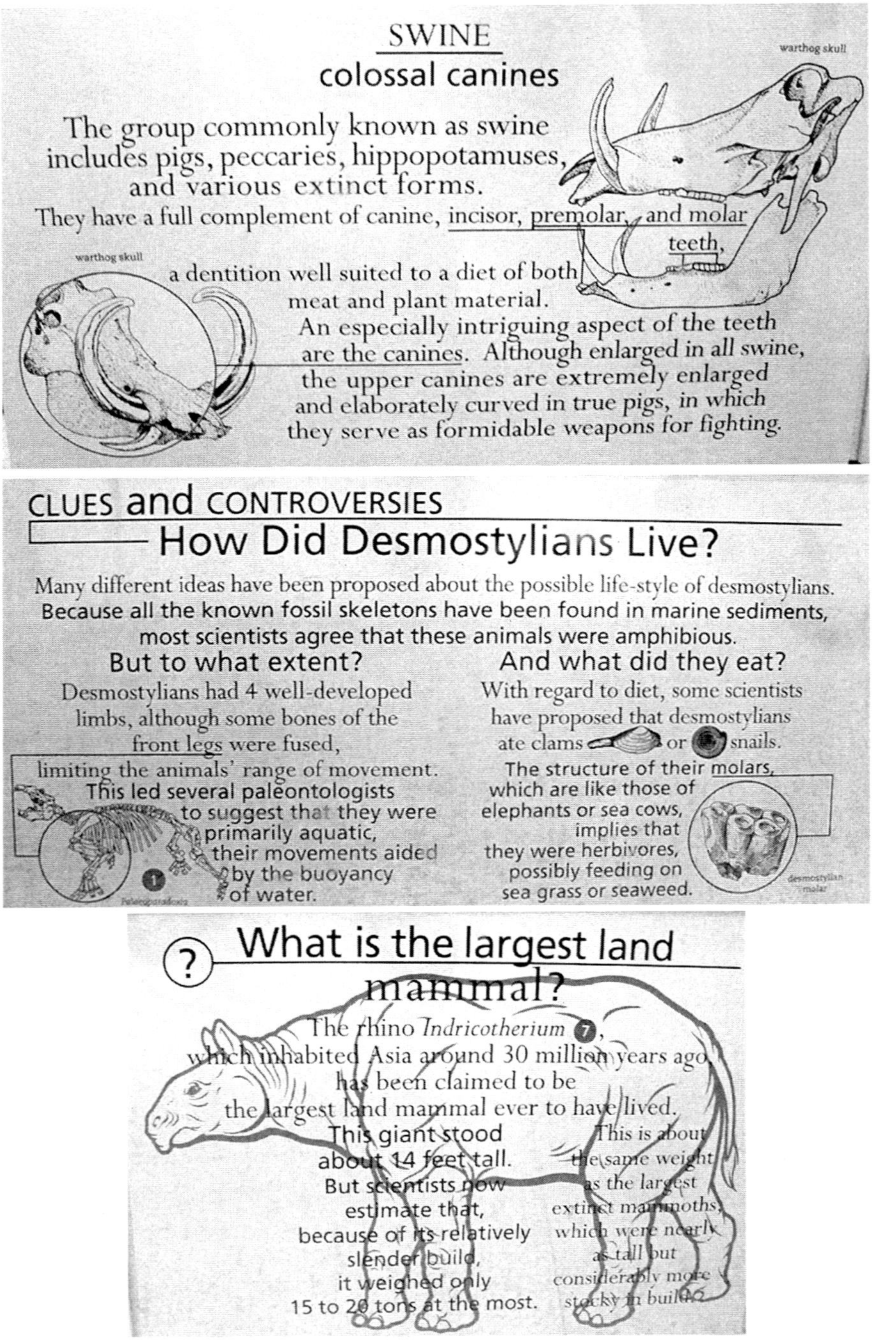

FIGURE 8.3. These panels from the *Paul and Irma Milstein Hall of Advanced Mammals* at the American Museum of Natural History feature centered text, underlining, inconsistent font usage, and awkward image placement, all resulting in a difficult reading experience for visitors. Photos by R. Kissel.

Images

In addition to text, images—photographs, original illustrations, or simple clip art—can also be used to convey information. As standalone elements or embedded in graphic panels and rails, they work together with general typography to convey a particular narrative and create the visitor experience. As Serrell (1996) suggests, the label text must be linked to what the visitor sees. Remember that all items within an exhibition, whether text, images, or illustrations, have content; all relay information and must work together to support the intent and goals of the exhibition.

Teams can be designed in such a way that the person writing the label copy is often not the same as the person designing the graphics. This approach does not suggest that the label copy would be written and then handed off to the graphic designer to import into panels. Instead, the team approach fosters an iterative process where the designer, developer, and content specialist work together to ensure that the text and images are greater than the sum of their parts. Together, these elements create the full visitor experience.

Images require more legwork beyond just selecting the "best fit." First and foremost, an institution must secure permission from the copyright holder of the image; do not assume that a non-profit institution using a particular image for educational purposes does not require permission. There can even, in some cases, be a fee assessed to secure rights, even for use in an exhibition. Attribution must also be considered at this point. Keep in mind that a source institution might have a very specific way they want an image from their collection cited, and this might be part of an image's negotiated use rights. Finally, before an image is selected, make sure it is of sufficient quality given the intended final output size—the actual final size of the displayed image. Otherwise it will result in a pixelated and poor quality print.[1]

Look and Feel

Images, objects, and text come together in space to create a complete physical experience for the visitor, often referred to as an exhibition's *look and feel*. It not only includes the general way things look, but also visitors' interaction with elements within a space and the impact that these components have on them. Look and feel includes obvious visual elements such as color, but also the variety of materials and finishes that, when taken collectively, create the experience of the space.

Consider the look and feel of the *James and Louise Temerty Galleries of the Age of Dinosaurs* at the Royal Ontario Museum (ROM) in Toronto versus the *Dinosaur Hall* at the Natural History Museum of Los Angeles County (NHM). The former opened in 2007 in the Daniel Libeskind designed Michael Lee-Chin Crystal, a glass, steel, and aluminum expansion of the original 1914 Italianate Neo-Romanesque building and 1933 neo-Byzantine addition.

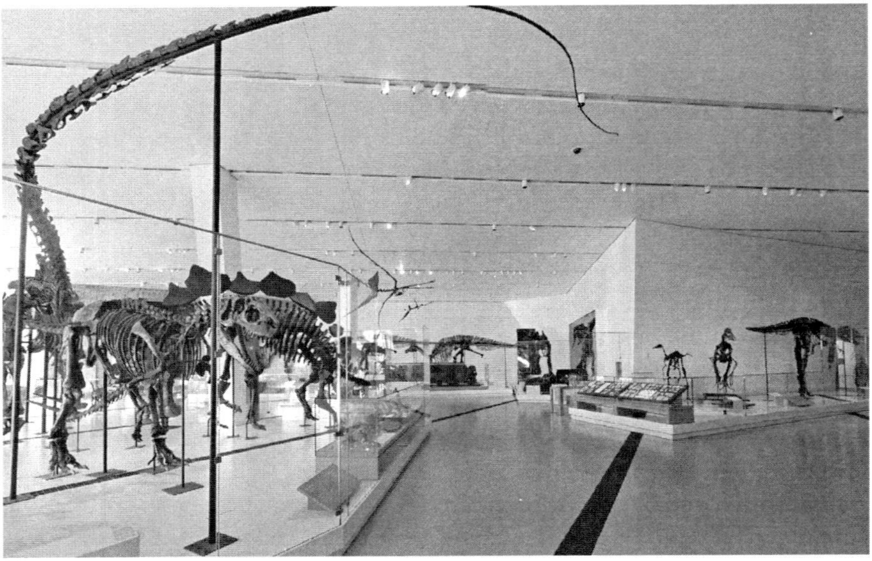

FIGURE 8.4. The permanent fossil galleries at the Royal Ontario Museum are set within the modern architectural frame of the Michael Lee-Chin Crystal. Photo by R. Kissel.

Twenty-five fully mounted dinosaur skeletons along with hundreds of additional fossils are displayed in open galleries under soaring 18-foot ceilings. The vast expanse of windowed walls of the Michael Lee-Chin Crystal offers vistas of the skeletons—including an impressive mounted *Barosaurus* (the only such mounted specimen in the world to include actual fossils)—to the world outside the museum. Large, brushed metal panels introduce the space and, together with the steel and glass of the Michael Lee-Chin Crystal and the stark white gallery walls, create a contemporary and slick feel to this display of cold ancient bones (Fig. 8.4).

Compare this aesthetic to the *Dinosaur Hall* at the Natural History Museum of Los Angeles County. The two-roomed, two-story atrium space spans the original 1913 building and the 1920s addition. The redesigned gallery, opened in 2012, includes vintage 1913 arched windows that bathe the space in natural light, much like the glass walls of the ROM's Michael Lee-Chin Crystal. However, in contrast to the sterile white walls and brief flashes of strong color (purple and blue) in ROM's exhibition, NHM's exhibition features orange, green, and deep blue tones that work together with cream walls to add both color and warmth to the space (Fig. 8.5). Central raised platforms offer vistas of large mounted skeletons without the obstruction of glass or plexiglass barriers. Visitors can walk around, and in some cases, under the skeletons—beyond even this, the second floor mezzanine allows visitors to see some of the large towering creatures eye-to-eye, thus creating an intimate visitor experience that encourages questioning and investigation, which are

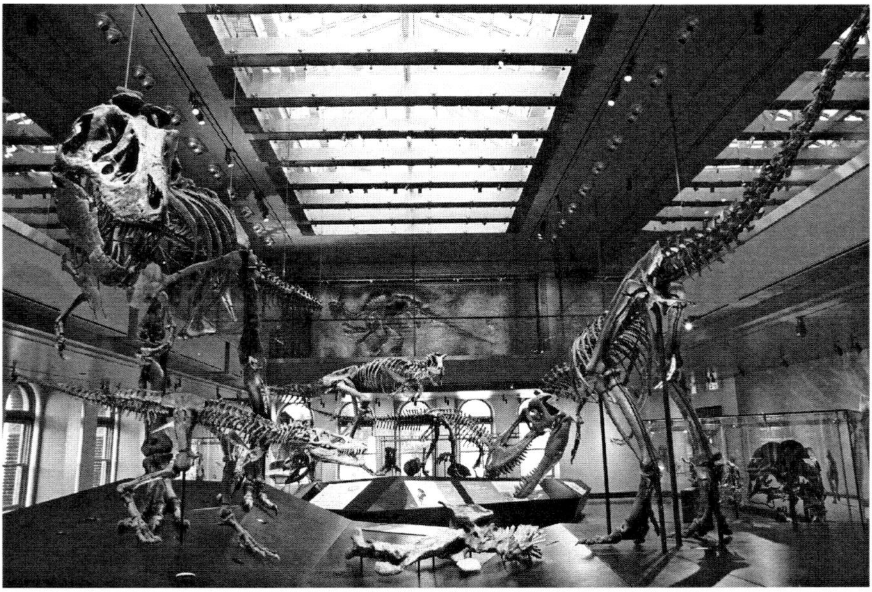

FIGURE 8.5. In contrast to those at the Royal Ontario Museum, the dinosaur skeletons at the Natural History Museum of Los Angeles County are surrounded by warmer colors and less obtrusive architecture. Courtesy Natural History Museum of Los Angeles County.

central to this exhibition's goals. These two galleries exhibit vastly different look and feels that contribute to two very different visitor experiences, in spite of the similarity of the objects displayed.

Color can be used to add cohesiveness to a particular exhibition, emphasizing and designating varied content within a space as well as working to evoke a general mood or ambiance. "Studies have shown that colour affects us physiologically, inducing calm or excitement, encouraging thought or stimulating imagination" (Maximea 2001, 171). Like text and typography, it can communicate content and work to advance or detract from an exhibition's visitor-related goals. Color can be used to evoke mood or emotion, or to emphasize and link content visually, drawing the eye to related themes and activities within a space. "Color...is such a powerful element, affecting all other aspects of the designed environment" (McLean 1993, 131).

In addition to the color embedded in the architectural space of a gallery, color gets incorporated into 2D graphics and must coordinate with an exhibition's overall typography. The basic decision comes down to light text on a dark background or dark text on a light background. Whichever is chosen, it is important to keep at least a 70% contrast between the foreground and background (Majewski n.d., 24). Also, keep in mind that reading light text on a dark background is generally more tiring to the eye (Serrell 1996), but it can work well when there is limited amount of body text or to grab a visitor's attention. "When using reverse type (light type against a dark background) sans serif may be a better choice...because dark backgrounds can diminish serifs by squeezing them" (Serrell 1996, 195). As a side note, there are certain color combinations that should be avoided when you consider overall accessibility; the inability to distinguish between red and green, for example, is the most common form of color-blindness.

As illustrated in the example of the dinosaur galleries at the ROM and NHM, color works in partnership with other materials to create the overall look and feel of the exhibition. Beyond a specific design intent, it is important that the team also be mindful of the ways in which visitors will interact with the galleries; materials contribute to both usability and visitor experience. Maximea (2001, 186) suggests that "the criteria for any material used for floor, walls, ceiling or fittings must include *how it will perform* [author's emphasis] within the physical setting of the gallery." This idea can be further broken down as follows:[2]

- *Appearance*—how materials fit within the overall look and feel of the exhibition.
- *Functionality*—"related to visitor comfort and safety, and efficient, flexible use of the gallery space" (Maximea 2001, 188). Certain materials can, for example, minimize visitor fatigue and encourage physical interaction in a space. You should maintain a contrast of at least 70% between the floors and walls in your space in order to increase perceptual differences and aid navigation (Majewski n.d., 46).
- *Maintenance and replacement*—considerations include potential visitor use, the ease and impact of cleaning, long-term availability, costs, and potential environmental impact.

Production

Navigating the ins and outs of graphic panel production can be a laborious process. The range of options, quality, and availability has increased substantially in just the last ten years, and it is likely to continue at a pretty brisk clip. Similar to the preceding discussion of typography, becoming familiar with some of the more common materials found in exhibitions will help the team navigate the process, and at least begin to become aware of the possibilities. So, even if you are working with outside designers and fabricators, understanding some of the basic options will help make decisions easier and create the opportunity for varied options to inform the process.

There are number of resources and industry groups that can point a team in the right direction. Trade show exhibition developers are often at the cutting edge of materials technology—corporate budgets spur innovation in materials and techniques from which museums with more modest exhibition budgets can benefit; this has been especially true, for example, in regard to the availability and affordability of green and sustainable materials options.

What follows is a list of common signage materials found in exhibitions. Although far from exhaustive, it presents some of the more common materials available.

- *Foam core* is a readily available expedient material, found standard in both black and white (though some colored options do exist). It is also available in acid free and self-adhesive varieties. It is made up of three layers: a polystyrene foam core sandwiched between two paper boards. Graphics can be printed directly on foam core or printed on other material and mounted to foam core with adhesive. It is inexpensive and offers simple, quick solutions for graphic panels.
- *Polystyrene* (styrene) is a versatile plastic material that is available in varying thicknesses. It can be cut, drilled, molded, bonded, silkscreen, and painted. Large graphics, such as murals, printed on thin styrene retain some flexibility, making them mountable to curved walls or other slightly irregular surfaces.
- Large format high-resolution display prints like Lambda or VUTEK produce sharp photographic resolution images. They can be wrapped around 1/16" signblank and adhered to a substrate (such as MDF or medium density fiberboard, a wood composite), a technique that is common and effective.
- *Vinyl lettering and graphics* printed on vinyl sheets can be used to make text and/or images adhere directly to a surface. They are available precut on sheets; based on the original digital files, the final sheets ensure correct spacing, eliminating the need to set each letter, word, or image individually.
- Often referred to as the Cadillac of outdoor signage, *porcelain (vitreous) enamel* can be quite expensive but offers a fabulous quality image, has a very long lifespan, and is equally at home indoors or outside.
- *High Pressure Laminate* (HPL) is also good for both indoor and outdoor applications, and is significantly more affordable than porcelain enamel while still offering good color quality and resolution. HPL is paper inkjet print sandwiched between multiple layers of melamine and phenolic sheets compressed at high pressure and temperature to form a solid core. The resulting graphic is a nice quality image that can be mounted, depending on the application, to a solid substrate like particleboard or MDF (medium-density fiberboard—wood fibers glued together under high heat and pressure). Bear in mind that in rare cases when there are natural or extreme fluctuations in ambient temperature, the natural ex-

pansion and contraction of a wood based substrate can cause adhesive failures. And remember to consider the relationship between the two materials you are mounting, for example, wood expands and contracts naturally, plastics do not. HPL can now be sourced with a post consumer recycled core.

- *Green and environmentally sustainable alternatives* are becoming more common as more and more natural history museums are increasingly aware of their responsibility as stewards of the environment. Sourcing green and sustainable alternative signage that incorporates post consumer materials, is fully recyclable, and/or incorporates non-toxic inks has become a viable and consistent alternative.

Conclusion

In a successful exhibition, the text, images, objects, and design all come together to support the Big Idea and exhibition goals. The 2D elements work in concert with the 3D to create a physical experience tied to narrative. There is a harmony between message and medium. Content and design work together to create the exhibition; they are experienced in tandem, and one cannot function without the other, and their sum is greater than their parts. The best content in the world cannot make up for bad design, and vice versa. Only in concert can they produce a meaningful and satisfying visitor experience.

Even in the absence of formal graphic training, it is important that team members understand basic design principals in order to leverage the concepts in support of the content. Whether it is decisions related to typography, layout, images, color, or signage materials, having an informed understanding of best practices offers a sound base from which to work. This brief introduction to the general ideas central to graphic and 2D exhibition design offers that base and promotes a productive process.

CHAPTER NINE

Hammer to Nail: 3D Design

"In all things of nature there is something of the marvelous."

—Aristotle

The actual physical space of an exhibition, the last major component presented in this book, works in concert with the exhibition's content and 2D design. Space is not a static medium, but is active in the creation and development of meaning. A visitor's interaction with space is a kinetic experience, engaging the whole body and accounting for movement through and within, and should be considered with the added fourth dimension of time. Time and space go hand in hand; it is a matter of not only knowing how a visitor interacts within the exhibition experience but also when and for how long. McLean (1993, 121) describes this as "pacing;" in other words, "the consideration of a visitor's movement through and interaction with an exhibition over time."

Tracking, Timing, and Museum Fatigue

Tracking and timing (T&T) studies have long been used to evaluate pacing and exhibition engagement. Discreetly following visitors around an exhibition and monitoring their interactions results in data that indicate the amount of time a visitor spends in an exhibition and with which elements they engage (Serrell 2010). T&T studies determine both the sweep rate (an exhibition's total square footage divided by the average time that tracked visitors spend in the exhibition) and the number of diligent visitors, or "the percentage of visitors in the tracked sample who stopped at more than one-half of the exhibit elements" (ibid., para 19). According to Serrell (2010), significant work on T&T since the 1990s has resulted in three major discoveries, the first of which suggests that visitors only spend an average of 20 minutes in any given museum exhibition. Serrell's earlier work suggests this can be broken down to an average sweep rate of 200 to 400 feet per minute (Serrell 1997), which means visitors are not spending a lot of time in any one exhibi-

Sarah J. Chicone and Richard A. Kissel, "Hammer to Nail: 3D Design" in *Dinosaurs and Dioramas*, pp 115-128.

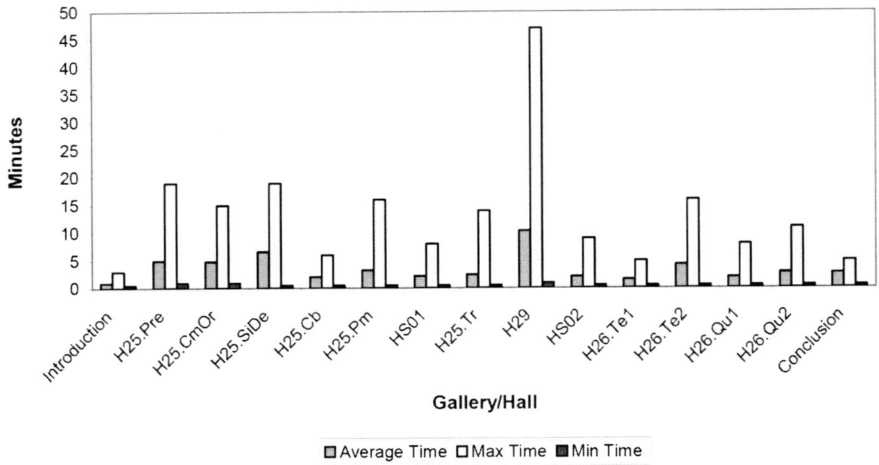

FIGURE 9.1. Part of the exhibition's summative evaluation, this Tracking and Timing study of *Evolving Planet* documents visitors' time within the exhibition's three main, similarly sized galleries. While the Precambrian & Paleozoic hall (H25), as well as the Cenozoic hall (H26), possess a single, directed path of exploration, the Mesozoic hall (H29) possesses a more open floor plan, allowing visitors more choice; for this reason, the Mesozoic displays were treated as a singular unit. Courtesy The Field Museum.

tion, considering the time and effort that an exhibition team devotes to the writing, design, and placement of a single label. Falk et al. (1985) and Falk (2009) present similarly challenging data, indicating that only 60 percent of a visitor's time within a museum was used for looking at displays, with the peak amount of content focus found in the first 15 minutes of a visit. Ultimately, there is a finite amount of time available to engage a visitor, so less is more. Serrell's second finding indicates that visitors explore smaller exhibitions in more detail, and her third indicates that exhibitions in science museums are often underused (Serrell, 2010, para 4). Figure 9.1 illustrates the results of the T&T study of the *Evolving Planet* exhibition at The Field Museum. Within a given institution there might be a number of related but separate exhibitions with a single, cohesive narrative, each made up of a variety of smaller displays that, when taken collectively, create the visitor experience. It is useful to determine which of these exhibitions are used most by visitors, and why, in any given institution. This information can then be translated into strategies that best engage a visitor with new content.

In light of tracking and timing studies, we want to also consider the impact of *museum fatigue*. First described by Gilman in 1916,[2] the phrase *museum fatigue* is generally used to illustrate the idea that "museum visitor interest towards exhibits decreased as visits progressed" (Davey 2005, 17). Davey (2005, 18–19) draws the following four general conclusions from the literature:

> 1) Interest reaches a high plateau for the first 30 minutes of a visit, and decreases thereafter...; 2) visitor interest decreases within smaller areas; 3) behavioral changes include cruising through galleries, relatively rapid rate of viewing without rest periods, and increased selectivity towards exhibits; and 4) patterns of fatigue are generally constant and predictable....

"Museum fatigue consists of several phenomena, and it seems likely that the causes of fatigue, and their importance, differ according to the scenario" (Davey (2005, 19), including both environmental and visitor attributes.

The findings from the T&T and museum-fatigue studies should be on the minds of every exhibition team, but the real question centers on the use of this information and how it can inform the approach to 3D design. Ultimately, these studies suggest that fewer and simpler elements within an exhibition will lead to increased usage.

Accessibility

Critical to the discussion of any public space is the consideration of accessibility. Accessibility for museums means "making the site's exhibits and programs available to all visitors" (Salmen 1998, 3). It is important to introduce the concept now; accessibility should not be an afterthought or a retrofit, but instead, a singular guiding principle for the entire approach to a space. (See also Chapters Six, Seven, and Eight for discussions on the accessibility of content, interactives, and 2D design.) The *Americans with Disabilities Act of 1990* (ADA) provided "for equal opportunity and independence for people with disabilities" (Salmen 1998, 8) and resulted in federal laws with accessibility standards that must be followed in given circumstances.[3] But beyond the legal mandates, it is simply good practice.

In talking about the general accessibility of museum exhibitions, it is useful to frame the conversation in terms of *Universal Design* (UD). UD is "the design of products and environments to be useable by all people, to the greatest extent possible, without the need for adaptation or specialized design" (Connell et al. 1997). In the "exhibits world" it can be thought of as "creating environments that benefit everyone, not just people with disabilities, and provide enjoyment and stimulation for all visitors" (Whitemyer 1997, 11). The concept is based on seven general principles:[3]

- equitable use
- flexibility in use
- simple and intuitive use
- perceptible information
- tolerance for error
- low physical effort
- size and space for approach of use

Tokar (2004, 10) suggests that, among science museum exhibition professionals, "universal design does not have a universal definition," but "within a broad range of definitions, the concept and practice... are well established.... Much remains to be done... but there is a firm basis for future success." We suggest that UD is more than just accommodating for disabilities; instead, it embraces designing for all and results in a mutual benefit for museum users—what Whitemyer (1997, 10) calls a "liberating ideal aimed at improving the quality of exhibits for everyone." The thread of accessibility and UD underlies all elements in this chapter. Now that we have it as the foundation, let's begin looking at space.

Moving through the Space

From the very beginning, an exhibition is faced with the task of grabbing the visitors' attention. An attractor, or an "attract" for short, assumes this key role. The "attract" is a visual or physical element that announces the exhibition and draws the visitor into a given space. In large institutions, there might be a number of exhibition experiences to choose from. The National Museum of Natural History in Washington DC, for example, has exhibitions on dinosaurs, human origins, the world's oceans, African art, and mammals, all on a single floor, and each of these exhibitions has its own comprehensive experience. Visitors must decide what to see and when to see it. A visitors' center at a nature preserve, on the other hand, may only have one exhibition comprised of a number of small, related display elements. But even then, visitors must decide if they are going to engage in the interpretive material or if they are going to simply enjoy the great outdoors; or, if both, then in what order? Careful consideration must be given to attracting visitors into a space. Perhaps it is a dynamic sign or a seminal object that first piques an interest (Fig. 9.2). But it is not enough just to bait visitors, you need to set the hook. Offer a straightforward introduction and orientation. Provide them with the information they need to decide whether they are willing to commit to the experience; answer the questions: What it is about? and How it will unfold?

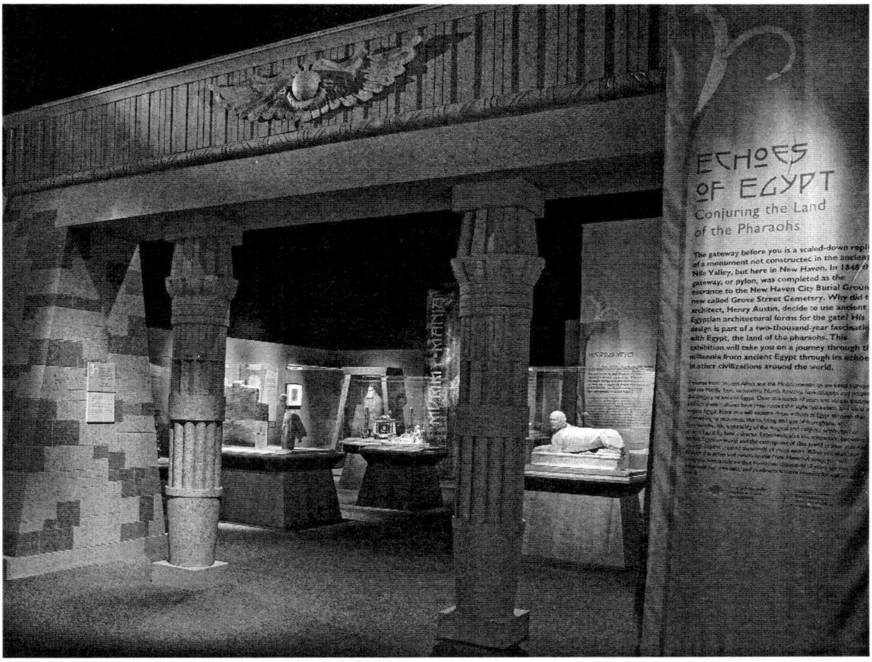

FIGURE 9.2. The entrance to the temporary exhibition *Echoes of Egypt* at the Yale Peabody Museum of Natural History offers a gateway to successfully pull visitors into the space, both physically and mentally. Photo by William Guth.

Once visitors are in an exhibition space, they need to be pulled through. This is where the content and design, both 2D and 3D, work together to lead visitors through the experience at a consistent pace. Traffic flow within the space impacts the sequence and tenor of the encounter, and designers must also consider the volume of visitors that must be accommodated at peak times (Maximea 2001, 104), with special attention given to potential bottlenecks within the space. Linear exhibitions offer consecutive experiences, each one dependent on the previous; this type of exhibition often takes the form of a chronological narrative. In natural history museums, this style is manifested in the ubiquitous march through geologic time—the Cambrian, to the Ordovician, to the Silurian, to the Devonian, and so on. A good example of this narrative can be found at the *Siegfried Family Hall of Ancient Life* at the Sam Noble Museum, or the Museum of the Earth's *A Journey Through Time,* both permanent galleries, where the exhibition's entrance is clearly marked and a chronological progression through sequential displays leads ultimately to the modern day and the exhibition's exit (Fig. 9.3). In contrast, a free form, open access exhibition encourages visitors to bounce around between exhibit elements, concepts, and objects to make their own connections, similar to the *Dinosaur Hall* at the Natural History Museum in Los Angeles. In these cases

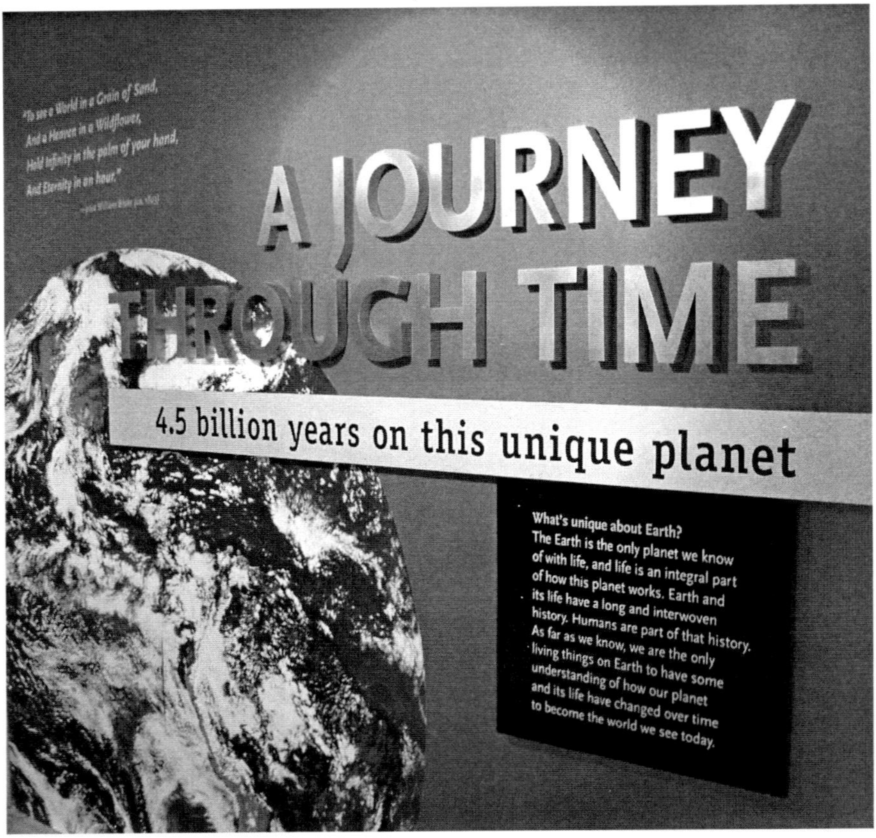

FIGURE 9.3. The entrance to Museum of the Earth's *A Journey Through Time* exhibition, which directs visitors into a chronological story told through sequential displays. Photo by R. Kissel.

inquiry or aesthetics can lead the visitor through the space; the experience is not dependent on consecutive events. Does the story pull visitors through, or does inquiry or aesthetics prompt further explanation and deeper engagement with the content? Which approach is used depends on the overall message or Big Idea, and the exhibition's experience and learning goals.

However, exhibition teams must never forget that "the best laid plans of mice and men often go awry." No matter the degree of planning devoted to a space, never discount the unexpected. Even within a highly structured linear experience, some visitors will interact with the exhibits in an unintended, non-linear way. It should never be assumed that all visitors will adhere to an intended path. Visitors might need to be guided visually through a space using attracts that engage their "sight lines" (McLean 1993), or through wayfinding techniques such as signs, maps, and other visual cues.

Whichever approach is taken, the accessible circulation route through an exhibition should be clearly marked and at least 36 inches wide for a one-way traffic flow, with a minimum width of 60 inches (although 72 inches is recommended) for a two-way traffic flow (Slamen 1998; Majewski n.d., 36). A route with one-way traffic flow should offer a passing space of 60 inches at least every 200 feet (ibid.). The route should also maintain a vertical clearance of 80 inches and be flat with a slope of no more that 5% (Majewski n.d., 39), with ramps or elevators provided to accommodate for any steps or stairs. Walls and floors should be well defined with a 70% contrast between vertical and horizontal surfaces (Majewski n.d., 41). Finally, do not underestimate the impact of sufficient seating: a summative evaluation of the *Gallery of California History* at the Oakland Museum of California (Randi Korn & Associates Inc. 2011) showed that visitors who used that exhibition's seating were likely to spend an average of 12 additional minutes engaged in the space compared to those that did not.

Lighting

Lighting is an important consideration for visitor experience; it works as an additional interpretative and curatorial tool (Shaw 2001), providing visual cues, creating mood, linking content, and highlighting, much in the same way that color and space do. Lighting impacts the way color looks in a space and where the eye is drawn. The *Evolving Planet* exhibition uses low lighting with blue undertones within the areas that present ancient sea life; together with an immersive 2D video wall, the lighting triggers an understanding within visitors that they are standing on the bottom of a prehistoric sea, even without reading a single label. This area is a stark contrast to the exhibition's dinosaur hall, where ambient light from exposed windows and strategically positioned spots highlight key features of the already impressive skeletal mounts.

TABLE 9.1. Light level recommendations for specific artifact and specimen categories

Examples	*Light levels*
Paper, textiles, feathers, dyed leather, felt	50 lux (5 footcandles)
All other organic materials	200 lux (19 footcandles)
Most stone, glass, ceramics, unpainted metal, and inorganic materials	300 lux (28 footcandles)

Source: Frost (2001, 119)

Beyond its potential as an interpretative tool, lighting should also be considered for its impact on general accessibility and object conservation (Table 9.1). This includes considerations of both natural and artificial light. A minimum of 10 footcandles (100 lux) is needed to adequately light an obstruction-free circulation path. LED lights are examples of SSL (solid state lighting) and are an alternative to traditional incandescent halogen lighting. They produce less ambient heat, which can have conservation impacts for sensitive objects. To prevent glare from shadows cast by visitors. It is recommended that objects and pictures be lit at a 30-degree aiming angle from above; tack height and distance from the wall can be determined based on an average viewer height of 5' 3" (Shaw 2001, 213).

Objects and Display

A given circulation route—whether linear or free-form—will take visitors past a variety of display elements, which together create the exhibition. These elements can include thematic groupings of interactives, immersive vignettes, graphics, text panels, and the all-important object.

Design plays a significant role in crafting visitor experience, and in creating a sense of authenticity. Museums are uniquely positioned to draw on the power of the authentic in their displays. This uniqueness and strength of experience can often be located within the objects themselves. Careful consideration of which objects are chosen and why, and how visitors engage with those objects within the space is required at the outset. That being said, a number of outside factors come into play—beyond the relevancy of a specific object—when weighing an object's inclusion within a particular exhibition. Authentic is ideal, but conservation ultimately rules when deciding between the real and a replica, and conditions that deter display of an actual object (e.g., too much light or humidity) are to be taken seriously. Part of the uniqueness of museums is that they present *the* opportunity to the public to see, engage, and interact with the actual, the real, the authentic. If visitors do not get to see a real *T. rex* in a museum, then where in the world would they ever get to see one? But whether the object is real, a cast, or a modeled replica, exhibition labels must clearly distinguish its classification in order to maintain a consistent level of trust with the visitor (Fig. 9.4).

How objects are displayed within a space has a significant impact on their potential engagement. The amount of space an object needs is called its "vista" and "relates to the comfort level of viewers either crowded too close to an object to view it comprehensively, or so distanced from the object as to not be able to view it clearly" (Maximea 2001, 82). For the most part, determining an object's vista is generally intuitive; large objects need more viewing distance, while small objects need less. A better view is offered when a visitor steps back from a mounted skeleton of a 70-foot-long *Apatosaurus,* while

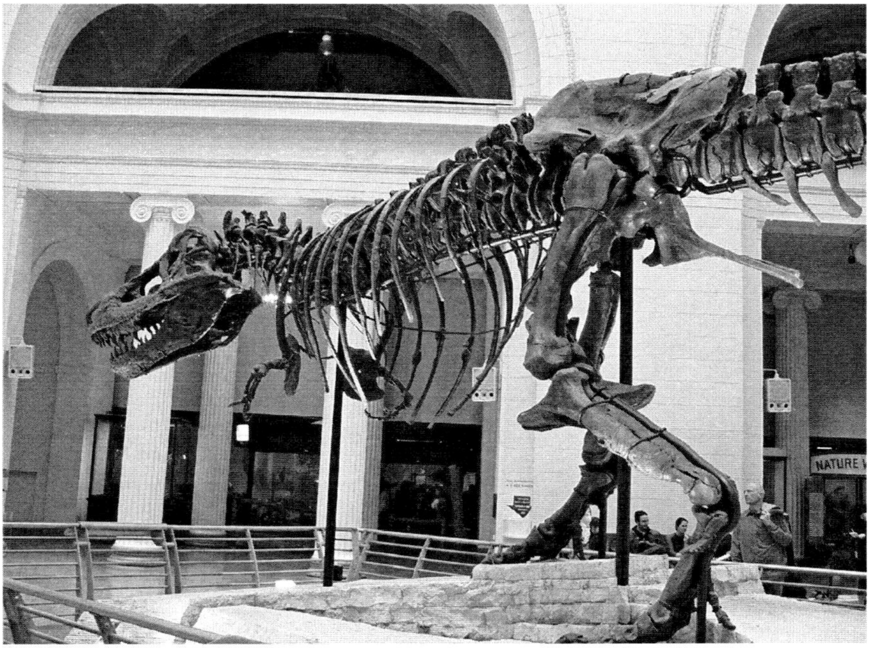

FIGURE 9.4. Sue is the largest, most complete and best-preserved *Tyrannosaurus rex* ever discovered. A visit to *Stanley Field Hall* at The Field Museum and you can see the real deal. Photo by R. Kissel.

getting up close to an Olduvai hand axe lets him or her see the striations and use wear. The number of objects in a given space relative to the suggested vista (amount of space needed to view the objects comfortably), combined with the circulation space needed to achieve the optimal vista, is considered the overall display density (Maximea 2001); different approaches and different objects require different display densities (Figs. 9.5). In general, visible storage—which places "entire collections at the disposal of the public for viewing and comparison" (Maximea, 2001, 149)—needs an estimated 50-60% of circulation space while thematic exhibitions need closer to 65-75% (Fig. 9.6).

Now that we have considered objects and the required amount of circulation room, let's move to thinking specifically about the means of display. The most common form of exhibition furniture is the "display case," or "vitrine" (Hahn 2001, 198) made up of a base with a transparent hood (usually either plexiglass[5] or glass) that protects an object. The base of the freestanding display case should be no more 36 inches above the floor (Salmen 1998; Majewski n.d., 44). This base might be either a solid plinth or a table. Freestanding table legs more than 12 inches apart, however, need to include a barrier easily detected by cane at no more than 27 inches above the floor (Salmen 1998, 79; Majewski n.d., 42). There should be a minimum of 60 inches (and when

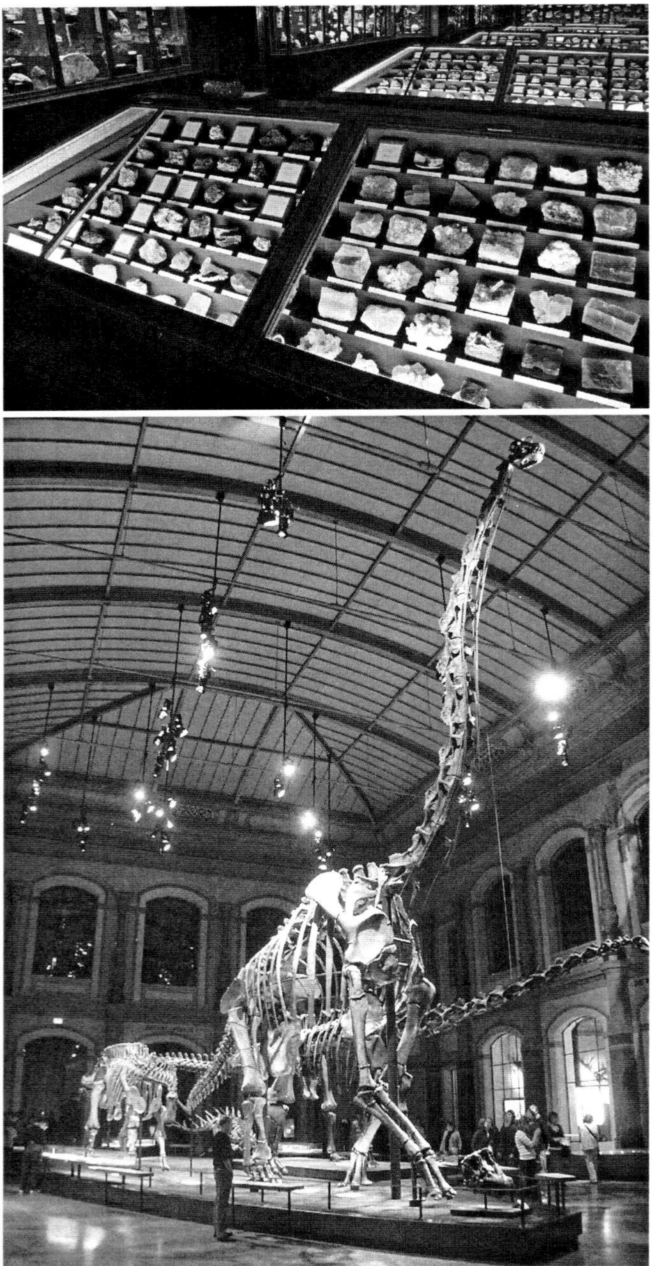

FIGURE 9.5. Smaller objects in this display of minerals at the Natural History Museum in Vienna are visible for close inspection, while large objects like this towering *Brachiosaurus* at Berlin's Museum für Naturkunde is viewed best from a distance. Photos by Jorge Royan (the display of minerals) and Axel Mauruszat (the *Brachiosaurus*), Wikimedia Commons.

possible 72 inches) between all cases, displays, and structures to allow for comfortable viewing. And U-turn routes around displays should be at least 48 inches wide (Salmen 1998). To accommodate viewing by visitors in wheel chairs, "each exhibit should have at least one 30-inch by 48-inch clear floor space located on each viewing side of the case or vitrine that is connected to or overlaps the accessible route" (ibid., 64).

Any object mounted to the wall, including artifacts, specimens, and cases, must not extend outward from the wall more than 4 inches, unless its bottom edge is lower than 27 inches above the floor or higher than 80 inches above the floor. Similar to free standing display cases, the base in a wall-mounted case should also be no more than 36 inches from the floor (Salmen 1998). In general, the shallower the display case, the easier the objects are to view. Larger objects should be mounted higher and smaller objects that require closer viewing should be mounted lower.

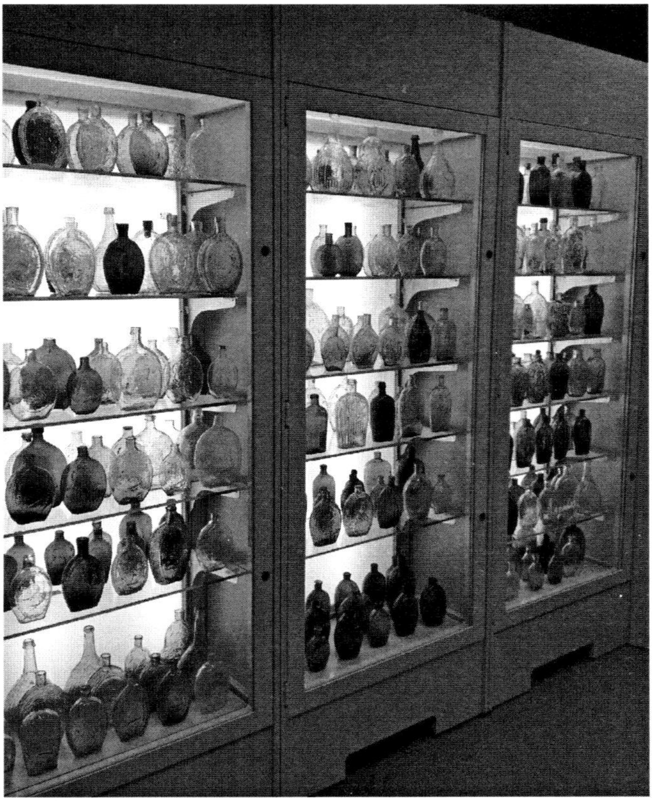

FIGURE 9.6. View of visible storage displayed in the Corning Museum of Glass. Developed in 1976, the visible storage found within the Museum of Anthropology at the University of British Columbia was the first use of this display technique. Photo by S. Chicone.

Within the cases themselves, multiple objects can be grouped together or a single object can be highlighted. Multiple items in a case should be varied in size in order to add dimension and visually attract the visitor's eye.

In addition to the environment external to an object case, careful consideration of the internal environment is just as important. Temperature, relative humidity, air movement, particulate matter and pollutants, pests, reactivity of material, and light can affect artifacts and specimens. These environmental factors should be considered for each individual specimen and artifact, as every object is unique and possesses unique curatorial needs. Requirements vary based on an object's material composition, and it is recommended that the optimal display conditions for all of the objects in an exhibition be determined sooner rather than later, as these conditions will likely affect decisions of design. Exhibitions, after all, are essentially an extension of an institution's collection management system. Dean (1994) and Frost (2001) provide an overview of these factors and methods for controlling the exhibition environment.

A Note on Organization

In the development of any exhibition, there can be 10s, 100s, or even 1000s of items to consider and manage, including but certainly not limited to graphic files, objects, photographs and images, media and hardware, and content references. The *Evolving Planet* exhibition at The Field Museum in Chicago had nearly 1200 specimens across its 27,000 square feet. The sheer volume of material requires a consistent and meticulous management philosophy applied to the entire project. There are a number of different database management programs that fit the bill, from Excel to FileMaker Pro, with varying degrees of utility and application. One way to maintain order is to assign numbers to the various exhibition elements. This allows the exhibition team to pinpoint specifics and keep track of progress. One example is a separate numbering system for each unique exhibition element that follows a three-signifier organization (xx-xx-xx).

The 1,200-square-foot traveling exhibition *Evolution on the Half Shell* will be used as an example to illustrate its utility. The first of the three signifiers marks the exhibition section. This particular exhibition was broken down into eight separate sections. Each section was given a discrete number:

- Introduction 01
- Biodiversity 02
- Biogeography 03
- Fieldwork 04
- Morphology 05

- Molecules 06
- Phylogeny 07
- Conclusion 08

In this example, 01-xx-xx would be anything found in the introductory section of the exhibition. The second signifier indicates the type of display element and varies depending on what strategies are employed in a given exhibition:

- G-graphic
- P-prop
- A-artifact
- S-specimen

In this example, 01-G-xx would indicate a specific graphic in the introductory section. The third signifier refers to a unique display element, which can be qualified even further; for example, all the graphic panels for the *Evolution on the Half Shell* exhibition had unique numbers with qualifiers as to their hierarchy:

- A panels: introductory or orientation panels
- B panels: section or group labels (secondary panels)
- C panels: rails
- D Panels: noninterpretive labels
- I panels: interactives

In this example, the introductory panel in the introduction section had the unique identifier 01-G-A01, while the section group labels were 01-G-B01, 01-G-B02, etc.

This system is useful in part because all elements within an exhibition can be organized in the same place—graphics, text, images, objects, etc. It is easy to locate them by their unique indicators and to sort out specific elements. Think about the organization of the exhibition content and space, and the best way to keep all the moving parts and pieces straight. Exhibition teams, regardless of exhibition size, need functional and applicable databases that work well with the development and design process.

Consideration of Budget

Finally, no discussion of design—or exhibition development in general—would be complete without a few words on budgets. The design and development process is not divorced from considerations of production and fabrication. Chapters Seven and Eight presented a host of physical considerations, including 2D and 3D elements, colors, cases, and materials. As the design and development process unfolds, teams must be mindful of project budgets at every step and every decision, as budget-induced constraints can significantly impact an exhibition's content delivery methods.

Budget navigation is not only time-consuming but it also requires ancillary research, pricing, and familiarity. Whether the project produces elements in-house, sources local talent, or contracts with a specialized exhibition design or fabrication firm will impact both cost and possibilities.

Regardless of the wide variety of options and pricing, teams should at least have a "sense" of what exhibitions cost. In a word, they are *expensive.* And in a time of financial instability, they can be some of the first things to get cut—in part or whole—from an institution's budget. So, how much do exhibitions cost? Well, it depends. But a general, working approximation is $150–$600 per square foot, a range that considers a diversity of exhibitions, from flat art to fully immersive. On occasion, per-square-foot costs are not just limited to physical fabrication of displays but can often include entire project costs, inclusive of design, research, fabrication, media, wall treatments, flooring, and lighting. It is important when working with outside consultants and designers to know what services and products they are including in their cost estimates; for example, design fees can easily run an additional 15 to 20 percent of fabrication costs. When considering budgets, don't forget about marketing, programming, operations, and other associated staff costs. Understanding your financial constraints at the beginning of your project will help relieve pressure on a team as it is forced to make difficult decisions during the planning process regarding what to keep and what to "value engineer." It is a good idea to budget a contingency amount at both the design and fabrication stages for any unexpected expenses or changes that might surface. If you have ever engaged in a home remolding project, after all, you are well aware that unexpected things can and often do happen; it is better to be prepared.

Conclusion

From thinking about the way visitors move and engage in a space to which objects to display and how to properly display them, an exhibition's 3D elements are as critical to the experience as the content and 2D design. Throughout all the considerations mentioned in this chapter—from lighting to organization—it is important to keep in mind the manifesto of UD and remember to reflect on the total visitor experience for all visitors. Considered together, the many elements of an exhibition will work in concert, resulting in a powerful, cohesive experience that is greater than the sum of its parts.

CHAPTER TEN

If You Build It, Will They Come? After the Opening...

"Somewhere, something incredible is waiting to be known."

—Carl Sagan

The exhibition has been installed, the final walkthrough complete, and the opening gala is in full swing. The exhibition team raises their glasses for a toast and basks in the glory of a completed project. Month after month of planning, long meetings, late nights, and hard work have led to this moment, as everyone breathes a sigh of relief. Every specimen is perfectly lit, and every label has been proofread and printed. The missing element? The most important element of the process: the visitor.

A Brief Note on Marketing

Considering the financial resources and the many hours, days, weeks, months, and in some cases years that can be devoted to an exhibition's development—from initial concept to public opening—reaching an audience represents the final, and perhaps most critical and difficult, challenge. That means bringing bodies and minds in the doors and into the gallery space. "*If you build it, they will come*" is a movie line, not a marketing strategy (Tonya Matthews, pers. com.). The Smithsonian Office of Policy and Analysis (2002b) suggests that museum marketing is more than just advertising (or waiting until *after* an exhibition has been developed, and then targeting efforts toward its promotion); instead, it means thinking broadly about market appeal—including visitors' wants and needs—at the very beginning and throughout the process. Museum marketing includes: 1) designing exhibitions to attract audiences; and 2) designing a promotion strategy to make visitors aware of said exhibitions (ibid., 3).

Though this discussion of exhibition marketing is near the conclusion of the book, we are in fact advocating a "marketing versus advertising" approach, with full consideration of audience at the beginning and throughout the process, not just after the exhibition has been developed. Full-scale marketing efforts begin with an understanding of current museum visitors and potential visitors, especially as they relate to a specific institution. What is the brand and identity of the institution, for example, and how does the exhibition fit into that brand?

Market analysis can be used to target specific sponsorships, or corporate donors, as well as determine invested parties and stakeholders, which ties back into Chapter Three's discussion of team members. Exhibition marketing can be approached from many different ways, including overall institutional branding, blockbuster experiences, Internet marketing including social media strategies, and traditional print media and ads (Office of Policy and Analysis 2002b). How an exhibition is marketed will depend on its intended audience and its content. Regardless of the approach, consider a full marketing strategy from the very beginning of an exhibition's development.

Measuring and Maintaining Success

Once the opening celebration is over, the members of the exhibition team return their focus to other projects that have been screaming for attention since the full court press to get everything installed and "perfect." So now what? The team walks away. Right? They have done their work, the exhibition has been installed, and nobody will shed a tear if they never again see another label about copepods or the Neolithic Revolution. Well, not exactly. Even though research, front-end and formative evaluation, and multiple rounds of prototypes have been used to ensure that the strategies employed do indeed meet the intended project goals, a lot can happen once exhibition installation occurs. How do the different elements work together in concert? Is the exhibition meeting its goals? While high visitorship rates are exciting, comforting, and welcome for many aspects of an institution, they are not a measure of an exhibition's effectiveness.

Enter summative evaluation. "Summative evaluation is [used] to assess the quality and impact of a fully implemented project" (Frechtling 2010, 7). Just because the last screw has been tightened does not mean that there is no longer room for changes. Some elements simply might not engage visitors, or they're confusing. While a few minor tweaks—rewording a label, moving a case—can be enough to fix the flow, certain elements might need significant reworking. Summative evaluation will provide the data needed to determine if the exhibition meets its stated goals.

Once the summative evaluation is complete, *now* everyone can walk away. Right? Of course not! Once the exhibition is installed, it is never a hands-off proposition; as covered in earlier chapters of this book, a tremendous amount of time and effort goes into developing and designing an exhibition, but a sustained effort is also needed to ensure that it continues to relay its message. Never underestimate the power of appropriate maintenance. Nothing frustrates visitors more than to have a litany of out-of-service display elements. Whether it is a wall of blank media screens or the blown projector bulb, the anticipation of use, wear, and expected maintenance should be considered during the design and development process, not *after* the projector bulb for an introductory video has blown and a yellow bus full of third graders has just arrived for their class tour. Maintenance includes:

- *Back Ups*, both physical and digital, for anticipated needs. Extra media players, projector bulbs, and monitors can be rotated in and out in a pinch, and digital copies of exhibition labels, media, and graphics can be updated, reinstalled, reprinted.
- *Operating Manuals* should include opening and closing instructions as well as troubleshooting tips for exhibition media. With a generally high turnover rate of museum staff, all manuals should be thorough and illustrated.
- *"As built" drawings* are images showing how elements were actually built. These illustrations are important, as elements can often change from the initial design.
- *General Appearance.* Don't forget to dust behind the mastodon skeleton. Dirty displays can leave a bad impression; if an institution doesn't take care of its objects or displays, it jeopardizes its credibility with visitors. General appearance also includes indicating when exhibitions are under repair, or objects have been removed.

In addition to the physical maintenance of an exhibition, also consider its academic maintenance. This consideration is particularly important in science-based exhibitions. As discussed in Chapter Two, our understandings about the world and our place in it are in a state of constant development. Scientific content has a short shelf life as new discoveries are made and scientific theories are refined over time. Being flexible and realizing the organic nature of content will help avoid headaches down the road.

Creating a Living Exhibition: Educational Programming

As shown in the word cloud in Chapter One, "education" is key to the mission of most, if not all, museums of natural history and science. Indeed, exhibitions are developed with the intent of informing visitors about the world and their place within it. On their own, exhibitions can instill awe and inspire productive curiosity. But why stop there? It is after the opening day that an exhibition truly comes to life, and much of that is born from not only the interaction of visitors with the exhibition's displays and elements but also the interaction of visitors with exhibition-specific educational programming.

In Chapter Three, we encouraged the inclusion of education staff in the development and design process. Fully embrace this opportunity. Were there ideas or concepts discussed in team meetings that did not find their way into the exhibition? Of course there were! Limited square footage and the resulting edits are a staple of the exhibition process. But these ideas can certainly be resurrected for educational programming. Whether presented as a lecture, part of a docent-led tour, or as activity sheets within a guide written for educators and their students, educational programming is an avenue that can and should be explored for expanding upon the material presented within the exhibition.

Involving educators in the development process can also provide critical pedagogical insight. The beauty of an exhibition team is the suite of talents found across its members, and rarely is there one member that possesses the talents of many. A great writer might not be an expert on the content, for example, and even the most scholarly of content specialists might not possess the knowledge and skill to teach their knowledge to museum audiences. Having an educator as part of the exhibition team can only add value to the team, helping to ensure that an exhibition is incorporating sound educational approaches. An educator can also help align the content to state or federal education standards; this practice can assist an exhibition greatly by justifying field trips for both traditional and home schools.

In summary, the involvement of educators during the exhibition development process will result in an exhibition that lives beyond the scope and interactivity of the physical displays. Avoid waiting until an exhibition's opening to ceremoniously pass the torch from the exhibition team to an institution's educators. If educators have been involved in appropriate team meetings throughout the process, they will have been able to contribute to the pedagogical approach of the exhibition. They will already be well informed about the exhibition's content and ideally have begun developing programming and activities well before opening day. As well, as many educators within a museum interact directly with visitors, they can train and prepare for exhibitions that might present potentially controversial topics such as evolution, energy, or climate. Such training is ideally led and supported by the museum as a whole, which can author a position statement that educators and all staff interacting with public can offer to any confrontational visitors.

Conclusion

An exhibition is rarely *finished*; instead, it is ready to open. And that opening is not the end of the process but simply one of the steps. After the opening, marketing efforts strive to reach potential visitors, and education staff bring the concepts and topics to life, beyond the static displays and objects. Because of this importance, we include marketing specialists and educators on our list of team players, and we recommend having them participate within the process from the onset or near to it. By keeping informed about and participating in the process, marketing staff can ensure that an exhibition's marketing matches the content, visual design, and tone of the exhibition. Educators can contribute to the pedagogy employed within the exhibition's presentation of content, and their early insight into the content and design of an exhibition will significantly help with the creation of educational programming.

Now Let's Take a Journey, Together

"There is a single light of science, and to brighten it anywhere is to brighten it everywhere."

—Isaac Asimov

On March 10, 2006, Chicago's Field Museum opened the doors to its permanent exhibition *Evolving Planet*. Contained within the space's 27,000 square feet is the nearly four-billion-year history of life's evolution on Earth, from tiny trilobites and towering dinosaurs to walking whales and massive mammoths. Along their journey, visitors find themselves surrounded by the otherworldly creatures of a 500-million-year-old sea, walking through an ancient Illinois forest that hums with the buzz of bird-sized insects, and standing face to face with chimp-like relatives from days long gone, exploring the roots of their very own family tree. Years in the making, *Evolving Planet* saw more than one million visitors within its first twelve months.

Invisible to the vast majority of visitors that explored *Evolving Planet*—and all exhibitions across the globe—are the endless decisions, disagreements, and details that lead to the exhibition's creation. Developing an exhibition is indeed a creative process. Whether the exhibition covers thousands of square feet or thirty, beginning with nothing more than a topic is an enormously daunting prospect. Every topic is bottomless, and every topic possesses endless avenues of exploration. Also overwhelming at first are thoughts of audience, tone, experience, and design—all considered within the shadow of looming deadlines and a rigid budget.

The key to the successful development of any exhibition is a well-organized, institutionally supported process. And teamwork. From the formation of an exhibition's overarching theme to the installation of the last object, the decisions of a team will ultimately shape the exhibition. Exhibition development—like the process of science—is subjective by nature, and this common thread connects all exhibitions and their displays. As a team sets out to deliver an exhibition, it wants to satisfy visitors' motivation-driven needs, as well as educate in an unbiased manner. But the development of an exhibition requires hundreds to thousands of decisions—decisions made by selected personnel. Ultimately, this means that no exhibition presents an objective point of view. Even the color chosen to serve as the backdrop for an object can influence interpretation. That color was one of countless decisions; it's simply a part of the process, but that decision can and will affect the visitors' experience. And while it's absolutely critical to have correct content and ground-breaking research presented within an exhibition's walls—indeed, it is a museum's responsibility to do so—it must also be accepted that the same topic developed independently by two or more institutions would likely produce very different exhibitions. Even if the overarching themes are similar, the writing, objects, design, and overall experience would vary tremendously.

Ultimately, then, exhibition development is not about the individuals' contributions. Instead, it is about the desire to teach and share with thousands—perhaps millions—of visitors a story of our truly amazing world. As a member of an exhibition development team, no matter the role, you are that teacher. To be such an influential educator is such a rare and unique gift. Embrace it and be humbled by it; your actions will influence countless lives and inspire many minds. Developing exhibitions is a great responsibility, and the process deserves respect. Those that participate in the process of exhibition creation should recognize that respect, and they should work closely with each other to create the best product possible. Always remember the visitor, as you are creating for them.

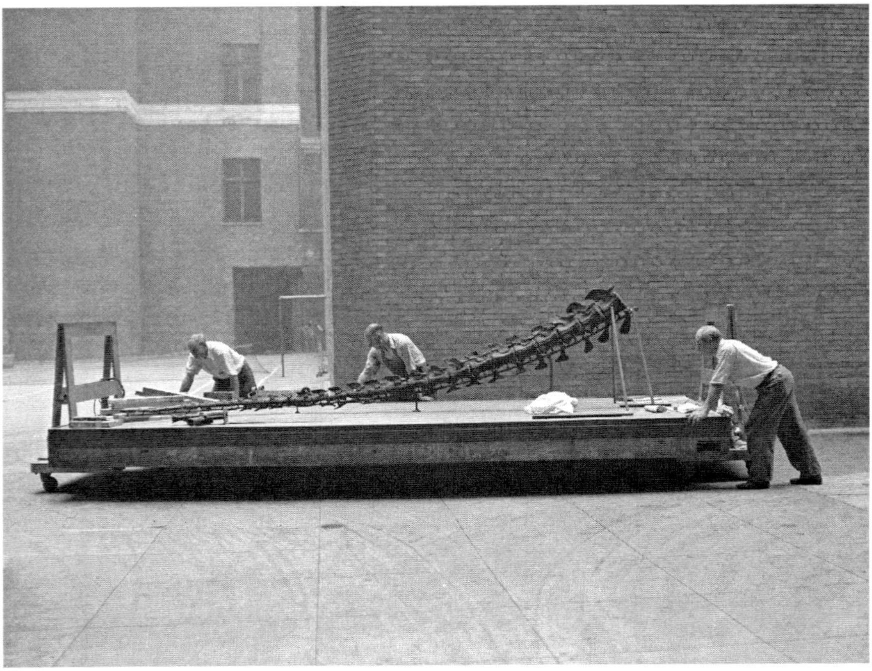

Staff from the American Museum of Natural History moving a tail section of "*Brontosaurus*" in June 1938. Image # 289651 American Museum of Natural History Library.

In closing, the exhibition design and development process is a difficult one, but the journey is both an enriching and rewarding experience. Embrace the challenge. And the responsibility. Nothing quite compares to the squeal of a seven-year-old child as she turns a corner to encounter a nearly 3,000-year-old mummy or the toothy grin of the fierce *Allosaurus*. It is easy to recognize that look of wonder and awe, as it is that very same look that beamed from your own seven-year-old self.

NOTES

PREFACE

1. Exhibitions, exhibits, and displays, oh my! The terms exhibition and exhibit are often synonymous within the museum field. The California Academy of Sciences uses exhibit, for example, while The Field Museum employs exhibition. For the purposes of this book, we prefer and use exhibition. An exhibition, then, is composed of smaller, individual displays.

CHAPTER 1

1. The term *specimen* refers to natural objects, such as fossils, skins, skeletons, and minerals, while human-made objects, such as projectile points, baskets, pottery, and sarcophagi, are termed *artifacts.*
2. Buffon strongly opposed and attacked Linnaeus' classification scheme, considering it too rigid to properly capture the fluidity of nature. To show his contempt, Linnaeus named a weed *Buffonia*.

CHAPTER 2

1. Designed for curricula within formal classroom settings, *A Framework for K-12 Science Education: Practices, Crosscutting Concepts, and Core Ideas*, published in 2012 by the National Academies Press, reinforces the definition of science as not only a body of knowledge but also the process of discovery and understanding; it also stresses the importance of interdisciplinary approaches.

CHAPTER 3

1. The team approach came on the scene with the "Kellogg Project." In 1979, the Kellogg Foundation funded the Exploratorium to conduct workshops for museum professionals, inclusive of the team approach to exhibition design (Office of Policy and Analysis 2002a, 12-13). For more information on the Kellogg Projects see Munley, M. E. 1986. *Catalysts for Change: The Kellogg Projects in Museum*

Education. Washington DC: The Kellogg Projects in Museum Education.

2. See Braden; Chiodo; Isble; Cohen-Stratyner; Rabineau; and Belew et al. in the Spring 2010 issue of the *Exhibitionist*, "Exhibition Frictions."
3. See the Spring 2007 special issue of *Exhibitionist*.
4. For more on indigenous perspectives for museums, see Susan Sleeper-Smith's 2009 *Contesting Knowledge: Museums and Indigenous Perspectives*, University of Nebraska Press.

CHAPTER 4

1. Until recently, the term hominid, or Hominidae, was used to describe the primate group that included the modern human species—*Homo sapiens*—and its extinct relatives, with the group Pongidae reserved for gorillas, chimpanzees, and the orangutan. But new information led to the discovery that the human family, gorillas, and chimps are more closely related to each other than any are to the orang. In light of this evidence, Hominidae now includes orangs, gorillas, chimps, and humans, with Hominini—or more simply hominin—reserved for the human family, past and present.

CHAPTER 7

1. Recall from Chapter One that Linnaeus' established convention for naming species of living things is a two-name system called binomial nomenclature. The first name refers to the genus, and the second is the species epithet. The generic name is capitalized, while the specific modifier is lowercase; both are always italicized. And both must be used when referring to a species. For example, *Tyrannosaurus* is a genus, which contains several, closely related species. *Tyrannosaurus rex* is one of those species. On its own, the name *rex* is *not* a species, and specific epithets should *never* be used on their own. Many biologists not only make this error but also insist that they are in the right.
2. This information is from "Words on the Walls: A Practical Guide to Museum Writing," an unpublished, though copyrighted, document by P. Rosenthal dated 2010.
3. As in Note 2.
4. As in Note 2.

CHAPTER 8

1. For a complete discussion of image resolution, we suggest consulting Parman, A., and J. J. Flowers. 2008. *Exhibit Makeovers: A Do-it-yourself Workbook for Small Museums*. Lanham MD: AltaMira Press (in particular, see p. 49).
2. This breakdown regarding materials is based on the three designations laid out by Maximea, H. 2001. Exhibition Galleries. *In* Lord, B., and G. D. Lord (eds), *The Manual of Museum Exhibitions* (pp. 143-185). Walnut Creek, CA: AltaMira Press.

CHAPTER 9

1. This generalization is based on Beverly Serrell's analysis of cumulative data from a 2009 Tracking and & Timing of summative evaluation in museum exhibitions, which contains 51 entries.
2. Gilman first described museum fatigue in 1916 in the article Museum Fatigue, published in *Scientific Monthly* 12:67-74.
3. This information is readily available at www.ada.gov, including ADA standards for accessible design. Other federal guidelines to be aware of include Section 508 of the *Rehabilitation Act* of *1973*, which deals with the accessibility of audiovisual displays (www.section508.gov), and the *Architectural Barriers Act of 1968*.
4. Version 2.0 – 4/1/97— http://www.ncsu.edu/project/design-projects/udi/center-for-universal-design/the-principles-of-universal-design/
5. Note that plexiglass requires special cleaner and should *not* be cleaned with glass cleaner.

REFERENCES

Adams, M., J. Luke, and T. Moussouri. 2004. Interactivity: Moving beyond terminology. *Curator* 47(2):155-170.

Adobe Systems Incorporated. 2000. *Typography Primer*. Adobe Systems Incorporated. Accessed January 10, 2010. www.adobe.com/education/pdf/type primer.pdf

Akerson, V. L., R. Abd-El-Khalick, and N. G. Lederman. 2000. Influence of a reflective activity-based approach on elementary teachers' conceptions of nature of science. *Journal of Research in Science Teaching* 37(4):295-317.

Alexander, E. P., and M. Alexander. 2008. *Museums in Motion: An Introduction to the History and Functions of Museums, Second Edition.* Lanham MD: AltaMira Press.

Alexander, V. 1996. Pictures at an exhibition: Conflicting pressures in museums and the display of art. *American Journal of Sociology* 101(4):797–839.

American Alliance of Museums. 2012. About museums: Facts. Accessed May 10 2013. www.aam-us.org/about-museums/facts

APE Project. (n.d.). In *Exploratorium*. Accessed May 10, 2013. www.exploratorium.edu/partner/ape/ape_intro.html

Attenborough, D. 2010. A Collection for the Future. *Museum of Life, Episode Six*: BBC Two. Viewed July 2012.

Beale, K. 2011. *Museums At Play: Games, Interaction and Learning.* Boston, MA: MuseumsEtc.

Belew, G., K. Gustafson-Hilton, S. Handy, and L. Wood. 2010. Splinters from green materials: Conversations about the frictions of green exhibition design. *Exhibitionist* 29(1):56-64.

Bell, P., and M. C. Linn. 2002. Beliefs about science: How does science instruction contribute? *In* Hofer, B., and P. Pintrich (eds), *Personal Epistemology: The Psychology of Beliefs about Knowledge and Knowing* (pp. 312-346). Mahwah, NJ: Lawrence Eribaum Associates.

Bell, P., B. Lewenstein, A. W. Shouse, and M. A. Feder. 2009. *Learning Science in Informal Environments: People, Places, and Pursuits.* Washington, DC: The National Academies Press.

Bine, J. 2006. What is a project manager... *Exhibitionist* 25(1):71-72.

Birchall, D., M. Henson, A. Burch, D. Evans, and K. Haley-Goldman. 2012. "Leveling Up: Towards Best Practice in Evaluating Museum Games." Paper presented at the annual conference of Museums and the Web, 11-12 April, San Diego. CA.

Bitgood, S. 1996. Evaluating the effectiveness of dioramas: A critical review. *Publics et Musees* 9:37-53.

Borun, M, and J. Dritsas, J. 1997. Developing family friendly exhibits. *Curator* 40(1):178-196.

Braden, D. R. 2010. Your personal toolkit: Easing through friction, fracas, and free-for-all. *Exhibitionist* 29(1):6-14

Bushara, L. 2009. Learning that play is learning. *Hand to Hand: Newsletter of the Association of Children's Museums*, Washington DC.

Catich, E. M. 1968. *The Origin of the Serif: Brush Writing & Roman Letters.* Davenport, Iowa: Catfish Press.

Chiodo, J. 2010. Putting our cards on the table: Discussing friction in the collaborative exhibition process. *Exhibitionist* 29(1):18-25

Clough, M. 2006. Learners' responses to the demands of conceptual change: Considerations for effective nature of science instruction. *Science Education* 15:463-494.

Cohen-Stratyner, B. 2010. Conflict in the house. *Exhibitionist* 29(1):18-25.

Conn, S. 1998. *Museums and American Intellectual Life, 1876-1926.* Chicago: University of Chicago Press.

Connell, B. R., M. Jones, R. Mace, J. Mueller, A. Mullick, E. Ostroff, J. Sanford, E. Steinfeld, M. Story, and G. Vanderheiden. The Principles of Universal Design. 1997. Version 2.0. Accessed July 10, 2013. ncsu.edu/ncsu/design/cud/about_ud/udprinciplestext.htm

Conniff, R. 2006. Happy birthday, Linnaeus. *Natural History* 115:42-47.

—. 2008. Buffon: Forgotten, yes, but happy birthday anyway. *The New York Times*, January 2.

Crean, B. 2001. Audio-visual hardware. *In* Lord, B., and G. Lord (eds), *The Manual of Museum Exhibitions* (pp. 214-222). Walnut Creek, CA: AltaMira Press.

Cronin, K. E., B. Stricker, R. A. Kissel, R. M. Ross, and C. Mundt. 2012. Ice, cold, animals: What museum visitors think about glaciers. *The Geological Society of America Meeting Exposition, Abstracts with Programs*: 448.

Csikszentmihalyi, M., and K. Hermanson. 1994. Instrinsic motivation in museums: why does one want to learn? *In* Hooper-Greenhill. E. (ed), *The Educational Role of the Museum, Second Edition* (pp. 146-160). London: Routledge.

Danks, M., M. Goodchild, K. Rodriguez-Echavarria, D. B. Arnold, and R. Griffiths. 2007. Interactive storytelling and gaming environments for museums: The interactive storytelling exhibition project. *In* Hui, K.-c., Z. Pan, R. C.-k. Chung, C.C.L. Wang, X. Jin, S. Göbel, and E. C-L. Li (eds), *Technologies for E-Learning and Digital Entertainment* (pp. 104-115). Berlin: Springer-Verlag.

Davey, G. 2005. What is Museum Fatigue? *Visitor Studies Today* 8(3): 17-21.

Dean, D. 1994. *Museum Exhibition: Theory and Practice.* New York, NY: Routledge.

De Queiroz, K., and J. Gauthier. 1992. Phylogenetic taxonomy. *Annual Review of Ecology, Evolution, and Systematics* 23:449-480.

Diamond, J., J. J. Luke, and D. H. Uttal. 2009. *Practical Evaluation Guide: Tool for Museums and Other Informal Educational Settings, Second Edition.* Lanham MD: AltaMira Press.

Dirbin, N. R., and A. Rickhoff. 2010. Media and the museum: A response to learning science in informal environments. *Curator* 53(2), 221-228.

Edwards, S. and D. T. Scahller. 2007. The name of the game: Museums and digital learning games. *In* Din, H., and P. Hecht (eds), *The Digital Museum: A Think Guide* (pp. 97-108). Washington, DC: American Association of Museums.

Evans, R. J. W. 1973. *Rudolf II and His World. A Study in Intellectual History 1576–1612.* Oxford: The Clarendon Press.

Fabrikant, G. 2012. Engaging children with the siren call of the app. *The New York Times*, October 26.

Falk, J. 2006. An identity-centered approach to understanding museum learning. *Curator* 49(2):151-166.

—. 2008. Calling all spiritual pilgrims! Identity in the museum experience. *Museum* 87(1):62-67.

—. 2009. *Identity and the Museum Visitor Experience.* Walnut Creek, CA: Left Coast Press, Inc.

Falk, J. H., and L. Dierking. 2000. *Learning from Museums: Visitor Experiences and the Making of Meaning.* Walnut Creek, CA: AltaMira Press.

—. 2002. *Lessons without Limits: How Free-choice Learning is Transforming Education.* Walnut Creek, CA: AltaMira Press.

—. 2010. The 95 percent solution. *American Scientist* 98:486-493.

Falk, J. H., L. D. Dierking, and M. Adams. 2006. Living in a learning society: Museums and free-choice learning. *In* Macdonald, S. (ed), *A Companion to Museum Studies* (pp. 323-339). Malden, MA: Blackwell Publishing.

Falk, J. H., J. J. Koran, L. D. Dierking, and L. Dreblow. 1985. Predicting visitor behavior. *Curator* 28(4):326-332.

Faron, R., and S. Curran. 2007. Team building: Thoughts on working well with others. *Exhibitionist* 26(1):32-38.

Findlen, P. 2003. The museum: Its classical etymology and Renaissance genealogy. *In*

Carbonell. B. M. (ed), *Museum Studies: An Anthology of Contexts, Second Edition* (pp. 23-45). London: Blackwell Publishing.

Frechtling, J. 2010. The 2010 User-Friendly Handbook for Project Evaluation. Directorate for Education and Human Resources: Division of Research and Learning in Formal and Informal Settings. National Science Foundation. Accessed July 15, 2013. www.westat.com/pdf/projects/2010ufhb.pdf

Fries-Gaither, J., and T. Shiverdecker. 2012. *Inquiring Scientists, Inquiring Readers: Using Nonfiction to Promote Science Literacy, Grades 3-5.* Arlington, VA: NSTA Press.

Frost, M. 2001. Preserving Collections in Exhibitions. *In* Lord, B., and G. D. Lord (eds), *The Manual of Museum Exhibitions* (pp. 110-123). Walnut Creek, CA: AltaMira Press.

Garcia-Luis, V., H. McDonald, and L. H. Migus. 2011. Multilingual interpretation in science centers and museums. Washington, DC: Association of Science-Technology Centers; and San Francisco: Exploratorium.

Grasso, H., and H. Morrison. 1999. Collaboration: Towards a more holistic design process. *Hooper-Greenhill, E. (ed), The Educational Role of the Museum* (pp. 172-177). New York: Routledge.

Gray, G., and V. B. Kendzia. 2009. Organizational self-censorship: Corporate sponsorship, nonprofit funding, and the education experience. *Canadian Review of Sociology* 46(2):161-177.

Hagan, H.A. 2008. The history of the origin and development of museums. In Genoways, H. H., and M. A. Andrei (eds), Museum Origins: Readings in Early Museum History and Philosophy (pp. 39-48). Walnut Creek, CA: Left Coast Press, Inc.

Hahn, T. 2001. Display cases. *In* Lord, B., and G. D. Lord (eds), *The Manual of Museum Exhibitions* (pp. 198-206). Walnut Creek, CA: AltaMira Press.

Haraway, D. 1990. *Primate Visions.* New York: Routledge.

Harrison, J. 2005. "What matters: Seeing the museum differently. *Museum Anthropology* 2(2):31-42.

Hayward, J. 2006. An evaluator is... *Exhibitionist* 25(1): 78-82.

Haywood, N., and P. Cairns. 2006. Engagement with an interactive museum exhibit. *In* McEwan, T., J. Gulliksen, and D. Benyon (eds), *People and Computers XIX – The Bigger Picture* (pp. 113-130). Berlin: Springer-Verlag.

Head, T. 2006. *Conversations with Carl Sagen.* Jackson, MS: University Press of Mississippi.

Hein, G. E. 1998. *Learning in the Museum*. New York: Routledge.

—. 2006. Museum education. *In* Macdonald, S. (ed), *A Companion to Museum Studies* (pp. 341-352). Malden, MA: Blackwell Publishing.

Hein, G. E., and M. Alexander. 1998. *Museums: Places of Learning.* Washington, DC: American Association of Museums.

Hoquet, T. 2010. History without time: Buffon's natural history as a nonmathematical physique. *Isis* 101(1): 30-61.

Hoyle, G. 2008. From theatrical illusion to ecological theater: The development of the classic wildlife diorama. *Journal of Natural Science Illustration* 40(8). Accessed July 10, 2012. www.gnsi.org/journal-article/theatrical-illusion-ecological-theater

Humphrey, T., and J. P. Gutwill. 2005. *Fostering Active Prolonged Engagement: The Art of Creating APE Exhibits. San Francisco, CA:* Exploratorium Museum Professionals Series.

Isble, M. 2010. Building and sustaining a culture of collaboration. *Exhibitionist* 29(1):26-35.

Janis, I. 1972. *Victims of Groupthink: A Psychological Study of Foreign-Policy Decisions and Fiascoes.* New York: Houghton Mifflin.

—. 1982. *Groupthink: Psychological Studies of Policy Decisions and Fiascoes.* New York: Houghton Mifflin.

Jennings, H. 2002. Go with the flow: The exhibit design process made visible. *Exhibitionist* 12(1):34-36.

Jennings, P. 2006. An exhibit developer is... *Exhibitionist* 25(1):73-75.

Johnson, L., H. Witchey, R. Smith, A. Levine and K. Haywood. 2010. *The 2010 Horizon Report: Museum Edition.* Austin, TX: The New Media Consortium.

Johnson, L., S. Adams Becker, H. Witchey, M. Cummins, V. Estrada, A. Freeman, and H. Ludgate. 2012. *The NMC Horizon Report: 2012 Museum Edition.* Austin, TX: The New Media Consortium.

Kamien, J. 2002. An advocate for everything: Exploring exhibit development models. *Curator* 44(1):114-128.

Kelly, L. 2007. Visitors and learning: Adult museum visitors' learning identities. *In* Knell, S. J., S. MacLeod, and S. Watson (eds), *Museum Revolutions: How Museums Change and Are Changed* (pp. 276-290). London: Routledge.

Kennedy, J. 1990. *User Friendly: Hands-on Exhibits That Work.* Washington, DC: Association of Science-Technology Centers.

Khishfe, R., and F. Abd-El-Khalick. 2002. Influence of explicit and reflective versus implicit inquiry-oriented instruction on sixth graders' views of nature of science. *Journal of Research in Science Teaching* 39(7):551-578.

Kinch, B. 2011. *The Functionality of Early Modern Collections.* Art and Design Thesis, Paper 69.

Knell, S. 2007. Museums, fossils and the cultural revolution of science: Mapping change in the politics of knowledge in early nineteenth-century Britain. *In Knell, S. J., S. MacLeod, and S. E. R. Watson (eds), Museum Revolutions: How Museums Change and Are Changed (pp. 28-47). New York:* Routledge.

Kohlsted, S. G. 1995. Museums: Revisiting sites in the history of the natural sciences. *Journal of the History of Biology* 28(1):151-166.

Koeppe, W. 2002. "Collecting for the Kunstkammer ". In *Heilbrunn Timeline of Art History*. New York: The Metropolitan Museum of Art. Accessed November 10, 2012. www.metmuseum.org/toah/hd/kuns/hd_kuns.htm

Kruse, J. 2005. Integrating the nature of science throughout the entire school year. *Iowa Science Teachers Journal* 35(2):15-20.

LaBar, W. 2009. The public Is a team member: Opening up the Cooking Exhibition's development process—in progress! *Exhibitionist* 28(1) 36-39.

Lederman, J. S. 2009. Teaching scientific inquiry: Exploration, directed, guided, and open-ended levels. National Geographic Learning, Cengage Learning.

Lederman, J. S., and N. G. Lederman. 2005. Nature of science is... *Science and Children* 43(2):53-54.

Lederman, N.G. 1999. Teachers' understanding of the nature of science and classroom practice: Factors that facilitate or impede the relationship. *Journal of Research in Science Teaching* 36(8): 916-929.

—. 2007. Nature of science: Past, present, and future. *In* Abell, S. K., and N. G. Lederman (eds), *Handbook of Research on Science Education* (pp. 831-880). Mahwah, NJ: Lawrence Eribaum Associates.

—. 2012. A powerful way to learn. *In* Froschauer, L. (ed), *A Year of Inquiry* (pp. 2-4). Arlington, VA: NSTA Press.

Lederman, N. G., and J. S. Lederman. 2004. Revising instruction to teach nature of science: Making the most of what you have. *The Science Teacher* 71(9):36-39.

Lindauer, M. A. 2007. Critical museum pedagogy and exhibition development. *In* Knell, S. J., S. MacLeod, and S. Watson (eds), *Museum Revolutions: How Museums Change and are Changed* (pp. 303-314). London: Routledge.

Lord, B., and G. D. Lord. 2001. *The Manual of Museum Exhibitions*. Walnut Creek, CA: AltaMira Press.

Lorentz, D. 2006. A study of the notions of immersive experience in museum based exhibitions. Masters Thesis: University of Technology, Sydney. Faculty of Design, Architecture & Building.

Lynn, N., J. P. Robinson, J. Patchen, and T. J. Tubutis. 2005. Old topic, new designs: Three new species of dinosaur exhibits. *Exhibitionist* 24(2):21-27.

MacDougall, B., and M. T. Carlson. 2009. West side stories: The blending of voice and representation through a shared curatorial practice. *In* Sleeper-Smith, S. (ed), *Contesting Knowledge: Museums and Indigenous Perspectives* (pp. 156-191). Omaha: University of Nebraska Press.

Majewski, J. n.d. Smithsonian Guidelines for Accessible Exhibition Design. Washington DC: Smithsonian Institution Accessibility Program. Accessed September 15, 2012. accessible.si.edu/pdf/Smithsonian%20Guidelines%20for%20accessible%20 design.pdf

Maximea, H. 2001. Exhibition galleries. *In* Lord, B., and G. D. Lord (eds), *The Manual of Museum Exhibitions* (pp. 143-185). Walnut Creek, CA: AltaMira Press.

Mazur, A. 2010. Do Americans believe modern Earth science? *Evolution: Education and Outreach* 3:629-632.

McLean, K. 1993. *Planning for People in Museum Exhibitions.* Washington, DC: Association of Science-Technology Centers.

Mitleer, R., K. R. Ginsburg, and Council on Communications and Media Committee on Psychosocial Aspects of Child and Family Health. 2007. The importance of play in promoting healthy child development and maintaining strong parent-child bond. *Pediatrics* 119(1):182.

Monroe, C., A. Lau, J. Schubel, and E. Cassano. 2006. Public Ocean Literacy: Making Ocean Science Understandable. *MCRI Aquatic Forum Report* 2006-2.

Montgomery, D. R. 2012. The evolution of creationism. *GSA Today: A Publication of The Geological Society of America* 22(11):4-9.

Morris, M. 2002. Recent trends in exhibition development. *Exhibitionist* 21(1):8-12.

Moser, S. 2010. The devil is in the detail: Museum displays and the creation of knowledge. *Museum Anthropology* 33(1):22-32.

Moss, D. M. 2001. Examining student considerations of the nature of science. *International Journal of Science Education* 23(8): 771-790.

Mouw, M., and D. Spock. 2007. Immersive media: Creating theatrical storytelling experiences. *In* Din, H., and P. Hecht (eds), *The Digital Museum: A Think Guide* (pp. 45-46). Washington, DC: American Association of Museums.

National Research Council. 1996. *National Science Education Standards.* Washington, DC: National Academy Press.

Newport, F. 2012. In U.S., 46% Hold Creationist View of Human Origins. Accessed September 1, 2012. www.gallup.com/poll/155003/Hold-Creationist-View-Human-Origins.aspx

Nicks, J. 2001. Curatorship in the exhibition planning process. *In* Lord, B. and G. D. Lord (eds), *The Manual of Museum Exhibitions* (pp. 345-372). Walnut Creek, CA: AltaMira Press.

Office of Policy and Analysis (OPA). 2002a. *The Making of Exhibitions: Purpose, Structure, Roles and Process.* Washington, DC: Smithsonian Institution, Office of Policy and Analysis.

—. 2002b. *Marketing Exhibitions: Will they Come?* Washington, DC: Smithsonian Institution, Office of Policy and Analysis.

Orseli, Paul. 2011. Prototyping 101. Accessed February 10, 2013. *blog.orselli.net/2011/03/prototyping-101-become-office-supply.html*

Padilla, M. 2012. Inquiry, process skills, and thinking in science. *In* Froschauer, L. (ed), *A Year of Inquiry* (pp. 24-26). Arlington, VA: NSTA Press.

Pallo, G. 2009. Genres of popular science: Urania and the scientific theatre. *In* Papanelopoulou, F., A. Nieto-Galan, E. Perdiguero, and M. Hernandez (eds), *Popularizing Science and Technology in the European Periphery, 1800-2000* (pp. 157-174). Burlington, VT: Ashgate Publishing.

Palmquist, S. 2011. Life Changes Summative Evaluation. New York Hall of Science.

Parman, A., and J. J. Flowers. 2008. *Exhibit Makeovers: A Do-it-yourself Workbook for Small Museums.* Lanham MD: AltaMira Press.

Perin, C. 1992. The communicative circle: Museums as communities. *In* Karp, I., C. M. Kreamer, and S. D. Lavine (eds), *Museums and Communities: The Politics of Public Culture* (pp. 182-220). Washington, DC: Smithsonian Institution Press.

Perry, D. L. 2012. *What Makes Learning Fun? Principals for the Design of Intrinsically Motivating Museum Exhibits.* Lanham MD: AltaMira Press.

Piacente, M. 2001. *Multimedia: Enhancing the experience. In* Lord, B., and G. D. Lord (eds), *The Manual of Museum Exhibitions* (pp. 222-233). Walnut Creek, CA: AltaMira Press.

Pitman, B. 1999. Muses, museums and memories. *Daedalus* 128(3): 1-31.

Plaza, C. 2009. In other words: Developing bilingual exhibitions. *ASTC Dimensions* July/August: 3-4.

Professional Networks Council of the American Alliance of Museums. 2012. Standards for Museum Exhibitions and Indicators of Excellence. Accessed May 10, 2013. www.name-aam.org/about/who-we-are/standards

Rabineau, P. 2010. Historical friction: Chicago perspective. *Exhibitionist* 29(1):51-55.

Rader, K, and V. Cain. 2008. From natural history to science: Display and the transformation of American museums of science and nature. *Museums and Society* 6(2):152-171.

Rand, J. 1993. Building on your ideas. *In* Bicknell, S., and G. Farmelo, *Museum Visitor Studies in the 90s.* (pp. 145-149). Science Museum: London.

Rand, J. T. 2009. Museums and indigenous perspectives on curatorial practice. I*n* Sleeper-Smith, S. (ed), *Contesting Knowledge: Museums and Indigenous Perspectives* (pp. 129-132). University of Nebraska Press, Omaha.

Randi Korn & Associates Inc. 2011. Summative Evaluation: Gallery of California History. [Randi Korn & Associates Inc.] Oakland Museum of California. Accessed July 12, 2013. www.informalscience.org/images/evaluation/2011_RKA_OMCA_HistoryGallery_summ_dist.pdf

Ravelli, L. J. 2006. *Museum Texts: Communication Frameworks.* New York: Routledge.

Rawson, L. R. 2006. Editorial and Graphic Guidelines. World Brooklyn Exhibition, Brooklyn's Children's Museum.

Reid, M., and B. Naylor. 2005. Three reasons to worry about museum researchers. *Museum Management and Curatorship* 20: 359-364.

Reiss, M. J., and S. Tunnicliffe. 2011. Dioramas as depictions of reality and opportunities for learning biology. *Curator* 54(4):447-459

Rodley, E. 2011. Gaming the Museum—Separating Fad from Function. Accessed July 7, 2011. exhibitdev.wordpress.com/2011/07/07/gaming-the-museum-separating-fad-from-function-part-four-of/

Roschelle, J. 1995. Learning in interactive environments: Prior knowledge and new experience. *In* Falk, J. H., and L. D. Dierking (eds), *Public Institutions for Personal Learning* (pp. 37-52). Washington, DC: American Association of Museums.

Rowlands, M. 2006. Presentation and politics. *In* Tilley, C., W. Keane, S. Kuchler, M. Rowlands, and P. Spyer (eds), *Handbook of Material Culture* (pp. 443-446). London, UK: Sage Publications.

Russick, J., and E. H. Nordstrom. 2010. Excellence in exhibition label writing competition 2010. *Museum*, September-October. Washington, DC: American Alliance of Museums.

Salmen, J. P. S. 1998. *Everyone's Welcome: The Americans with Disabilities Act and Museums.* Washington, DC: American Association of Museums.

Samis, P. 2007. Visual velcro: Hooking the visitor. *Museum*, November/December.

Sampson, S. D., and S. George. 2004. Reinventing a Natural History Museum for the 21st Century. *Proceeding of the California Academy of Sciences* 55(13): 283-294.

Schwartz, R. S., N. G. Lederman, and B. Crawford. 2004. Developing views of nature of science in an authentic context: An explicit approach to bridging the gap between nature of science and scientific inquiry. *Science Education* 88(4): 610-645.

Serrell, B. 1996. Exhibit Labels: An Interpretive Approach. Walnut Creek, CA: AltaMira Press.

—. 1997. Paying attention: The Duration and Allocation of Visitors' Time in Museum Exhibitions. *Curator* 40(2):108-113.

—. 2010. Paying More Attention to Paying Attention. *In* Center for Advancement of Informal Science Education (CAISE) Newsletter. Accessed July 10, 2013. www.caise.insci.org/news/96/51/Paying-More-Attention-to-Paying-Attention/d,resources-page-item-detail

Shaw, K. 2001. Lighting. *In* Lord, B., and G. D. Lord (eds), *The Manual of Museum Exhibitions* (pp. 207-213). Walnut Creek, CA: AltaMira Press.

Shelton, A. A. 2006. Museums and museum displays. *In* Tilley, C., W. Keane, S. Kuchler, M. Rowlands, and P. Spyer (eds), *Handbook of Material Culture* (pp. 480-499). London, UK: Sage Publications.

Silverman, L. H. 1999. Meaning making matters: communication, consequences, and exhibit design. *Exhibitionist* 18(2):9-14.

—. 2010. Visitor meaning-making in museums for a new age. *Curator: The Museum Journal* 38:161-170.

Simon, N. 2009. Design for participation. *Exhibitionist* 28(2):12-17.

—. 2010. *The Participatory Museum.* Santa Cruz, CA: Museum 2.0.

Sims, J. 2006. An exhibit designer is..." *Exhibitionist* 25(1):76-77.

Sleeper-Smith, S. 2009. Contesting knowledge: Museums and indigenous perspectives. *In* Sleeper-Smith, S. (ed), *Contesting Knowledge: Museums and Indigenous Perspectives* (pp. 156-191). Omaha: University of Nebraska Press.

Sloan, P. R. 1976. The Buffon-Linnaeus controversy. *Isis* 67(3):356-375.

Spary, E. C. 2000. *Utopia's Garden: French Natural History from Old Regime to Revolution.* Chicago: University of Chicago Press.

Spencer, H. A. D. 2001. Exhibition Text Guidelines. *In* Lord, B., and G. D. Lord (eds), *The Manual of Museum Exhibitions* (pp. 398-399). Walnut Creek, CA: AltaMira Press.

Spock, D. 2004. Is it interactive yet?" *Curator* 47(4): 369-374.

Stein, R. 2012. Chiming in on museums and participatory culture. *Curator* 55(2):215-226.

Steiner, C. 1991. Worlds in miniature, worlds apart: Dioramas, models, and mannequins in Peabody Museum exhibitions, Peabody Museum, Harvard University. *Museum Anthropology* 15(2):34–37.

Taquet, P. 2007. Establishing the Paradigmatic Museum: George Cuvier's Cabinet D'Anatomie Comparee in Paris. *In Knell, S. J., S. MacLeod, and S. E. R. Watson (eds), Museum Revolutions: How Museums Change and Are Changed (pp. 3-14). New York:* Routledge.

Terrell, J. 1991. Disneyland and the future of museum anthropology. *American Anthropologist* 93:149-153.

Tokar, S. 2004. Universal design in North American museums with hands-on science exhibits: A survey. *Visitor Studies Today* 7(3):7-10.

Weil, S. 1999. From being about something to being for somebody: The ongoing transformation of the American museum. *Daedalus* 128:229-258.

Whitemyer, D. 1997. Creatively challenged: Toward universal design. *Exhibitionist* 16(2);10-12.

Wyman, B., S. Smith, D. Meyers, and M. Godfrey. 2011. Digital storytelling in museums: Observations and best practices. *Curator* 54(4):461-468.

INDEX

NOTE: Italicized numbers refer to figures & figure captions; if a topic or institution is found within both the body and figure on the same page, the number is listed twice—once set in roman and once italicized.

ABOUT THE AUTHORS

Dr. Sarah J. Chicone is Senior Lecturer and Assistant Director of the Museum Studies Graduate Program at Johns Hopkins University. She has been lead curator, developer, and designer for contemporary art, history, anthropology, and natural history exhibitions. Sarah has a breadth of experience in both large and small exhibition projects, and has worked in a variety of capacities that range from Director of Exhibits for a small natural history museum to a content developer/ coordinator for an exhibition design firm. As an archaeologist, museologist, and educator, her academic and professional interests include material culture studies, informal education, cultural heritage, exhibition design and display, and global museum initiatives and partnerships, as well as the archaeology of working class poverty, and public archaeology.

Dr. Richard A. Kissel is Director of Public Programs at the Yale Peabody Museum of Natural History, where he leads public education and the development, design, and production of exhibitions. He is a vertebrate paleontologist, science educator, and author with more than 20 years of experience within the museum field. From the deserts of Texas to the forests of Germany, he has traveled the globe to excavate and study the remains of dinosaurs and other ancient beasts. Richard is a featured scientist online at NOVA's scienceNOW, and he was the lead curator and a developer for *Evolving Planet*—The Field Museum's 27,000-square-foot exhibition on the history of the Earth and its life. Richard has authored scientific articles, popular pieces, and children's books on paleontology and the nature of science, and his artwork has been featured in exhibitions, scientific and popular articles, and the New York Post. He is also an instructor for the Museum Studies Graduate Program at Johns Hopkins University.